QUANTUM PHYSICS

FOR BEGINNERS

AN EASY GUIDE FOR DISCOVERING THE HIDDEN SIDE OF REALITY ONE SPECK AT A TIME

TABLE OF CONTENTS

DISCLAIMER

THE FOLLOWING MANUAL ONLY CLAIMS TO BE EXHAUSTIVE IN SOME OF THE BREADTH AND FACETS OF THE SUBJECT. SOME OF THE EXAMPLES ILLUSTRATED ARE THE MOST KNOWN AND EASIEST TO GRASP IN THE FIELD AND A MYRIAD OF OTHER EXAMPLES AND CASES HAVE BEEN DELIBERATELY DISCARDED IN ORDER TO MAKE THIS TEXT THE MOST ACCESSIBLE TO EVERYONE REGARDLESS OF THEIR LEVEL OF STUDY.

THEREFORE, THE TEXT MAY BE REDUCTIVE AND SIMPLISTIC IN MANY OF ITS PARTS. FOR THOSE WHO ALREADY HAVE A GOOD LEVEL OF KNOWLEDGE OF PHYSICS, OTHER TYPES OF TEXTS ARE RECOMMENDED.

ALL EFFORTS HAVE BEEN MADE TO MAKE THE CONTENT ACCESSIBLE AND ENJOYABLE BY PURGING IT OF COMPLEX MATHEMATICS AND FORMULAS (WHICH WOULD REQUIRE GREATER PREPARATION TO UNDERSTAND THEM) WHILE MAINTAINING THE KEY ESSENTIAL CONCEPTS.

WELCOME

Welcome to Quantum Physics for Beginners!

Does the thought of learning quantum physics make you want to curl up in a ball and cry? Then you have come to the right place because we're here to wipe away those tears and turn that frown upside down. We're going to take a real-world, relatable look at the science behind the tiniest pieces of our universe and the scientists who study them. Part of the reason quantum physics is so intimidating is that the subject is often fraught with indecipherable formulas and fancy terminology that leaves the average human scratching their head and throwing up their hands.

So go ahead and use those stuffy textbooks to prop up the uneven legs on the table you inherited from Aunt Sarah and Uncle Joe.

Sure, we're going to have to tell you about some of the math, but we promise it won't be painful.

Even better, there won't be a test at the end. Just sit back, relax, and come on a laid-back, sometimes irreverent ride that will take you from the discovery of the atom to the astonishing science of the greater universe.

We're going to take a look at some of the most brilliant minds that ever grace a laboratory, the experiments they developed and presented to the world, and why sometimes, even these geniuses were misguided in their pursuits.

Quantum physics strives to recognize and explain how the universe's tiniest pieces all fit into the big picture. If there comes a time when human beings can journey to and live on other worlds, it's going to be because of quantum physics. If we ever want to travel through time, it's going to be because of quantum physics. And if you have no interest in those things, but you'd like a teleporter to get your errands done a little quicker- you guessed it-it'll take some pretty solid quantum physics. Yes, these things all sound more like science fiction than science fact, but have you ever read any good ol' pulp sci-fi? Some of that stuff has already come true, so there's always hope!

The point we're trying to make is that quantum physics isn't something you need to be afraid of.

As you work your way through the book, you'll realize that the scientists who founded and continue to study quantum physics might be wired a little differently (looking at you, Schrödinger!)

Still, they're human beings who want the same thing as you, to know and understand the nature of the world and the vast universe around us.

INTRODUCTION TO QUANTUM

Scientists didn't just wake up one day with a laboratory full of fancy electron-scanning microscopes and mass spectrometers. It took years of discovery and experimentation to get to the understanding of quantum physics that we have today.

The story of quantum physics, which holds a dictionary definition of *"the branch of physics concerned with the quantum theory"* (helpful, huh?), begins with the story of classical physics. This, of course, is the scientific discipline that aims to use mathematics and mechanics to explain how the world works. So, where does the fundamental split between classical and quantum physics occur?

The truth is, there is no specific date or definitive line that shows us where classical physics ends and quantum theory begins. Like a snowball rolling downhill, the theories that would evolve into quantum studies started slowly, building up momentum and growing - and that snowball eventually became an avalanche that is still roaring today.

It was innate human curiosity that instigated the first experiments that would lead eventually to the creation of quantum physics as a separate field of study. When scientists found that they couldn't explain things using the laws of classical physics, they went looking for their answers elsewhere. Thus, quantum physics, quantum mechanics, quantum chemistry, and every other quantum discipline was born.

But let's get things in the right order.

1.1 What is Quantum?

At its core, quantum refers to the fundamental nature of matter and energy at the smallest scales. It encompasses the behavior and interactions of particles of the subatomic world. It is derived from the Latin word "*quantus,*" meaning "how much" and/or "how many."

The term "**quantum**" is employed to express the fundamental basic units or packets of energy, that exhibit both particle and wave-like properties.

The term "quantum" encapsulates the notion that these physical quantities exist in discrete, quantized states, rather than in continuous, infinitely divisible forms. It signifies a departure from classical physics, where quantities were assumed to vary continuously and could take any value.

Quantum measures vs Classical measures

A quantum measure and a continuous and infinite measure are two distinct concepts that differ in terms of their nature and application.

A quantum measure refers to a measurement that is discrete and discontinuous in nature. It is characterized by distinct and separate values or states. To make it simpler (and may physicists forgive this profane example): let's consider a ruler, a "*quantum ruler*" - a quantum measure would involve measuring an object's length using discrete units, such as centimeters or inches in this fictitious case.

Each unit represents a specific value, and there is no possibility of subdividing the unit further. For example, if this hypothetical quantum ruler has centimeter markings, you can measure an object only as 1 cm, 2 cm, 3 cm, and so on. It is not able to give you a value below that unit. There is no room for any intermediate values between these discrete units.

On the other hand, a continuous and infinite measure refers to a measurement that is not limited to discrete unit "blocks" but can be subdivided infinitely. Using the ruler example, a continuous and infinite measure would involve measuring an object's length with infinite precision. Each centimeter on the ruler can be further divided into an infinite number of smaller units, such as millimeters or micrometers, and so on. This means that between each centimeter marking on the ruler, there are countless intermediate values that can be measured.

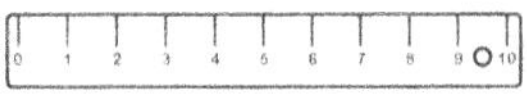

INDIVISIBLE UNITS IN THE SUBATOMIC WORLD

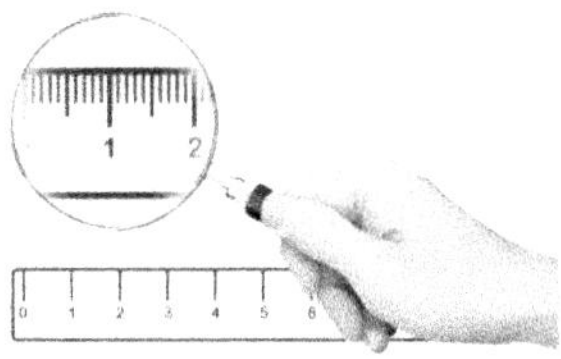

CONTINUOUS AND INFINITE DIVIDISIBLE UNITS

In summary, a quantum measure involves discrete and separate values, while a continuous and infinite measure allows for infinite subdivisions and intermediate values.

The infinite subdivision is a pure mathematical possibility that only makes sense on paper. But in reality, things are a little different.

The revolution of quantum physics brings to the table this new discovery, which is: at the subatomic level there's a limit in which the matter and energy can be divided. **Matter and energy cannot be divided infinitely.** Understanding the difference between these two types of measurements is crucial in quantum physics.

Said that, today, the term "quantum" has extended beyond the realm of physics and permeated various fields, including computing, cryptography, and even spirituality. It has become synonymous with notions of uncertainty, unpredictability, and the potential for revolutionary breakthroughs.

1.2 How it was born

Even before the apple could fall on Newton's head and spark the origin of classical physics, quantum physics made a brief debut.

The earliest concept of little tiny pieces of matter was actually introduced circa 400 BCE by the Greek philosopher Democritus. He proposed a so-called *Theory of the Universe* that subscribed to the following principles:

- all matter consists of indivisible spherical pieces that he called atoms; these are too small to be seen;
- there is empty space between all atoms;
- atoms are solid but have no set internal structure;
- every substance is made of different atoms, with their own weight, size, and shape.

This early theory is surprisingly accurate, albeit generalized. The fact that Democritus based his theories on philosophical thought experiments rather than scientific study makes his theory's accuracy even more astounding.

He firmly believed that all substances could be divided and divided again until only the smallest elements of that matter remained - and that smallest particle is what he dubbed "**the atom**." In Greek, the word "*atomos*" translates to "indivisible" or "uncuttable," which perfectly encapsulates this concept.

His theory would remain widely accepted and largely untouched until the turn of the 19th century.

As mentioned, there is no precise date in the physics timeline that signals the beginning of this new branch but it emerged as a distinct discipline when scientists started to question the principles of classical physics and sought new explanations for the behavior of matter and energy. This quest for knowledge led to groundbreaking discoveries and a paradigm shift in our understanding of the fundamental nature of reality.

One of the key figures in the development of quantum physics was Max Planck, who introduced the concept of quantization and the idea that energy is quantized into discrete packets called '**quanta**' (plural of quantum). This revolutionary concept laid the foundation for the field and set the stage for further advancements.

As scientists delved deeper into the study of matter and energy, they encountered phenomena that classical physics could not explain, such as the behavior of light, and the fact that particles can exist in multiple states and can be in multiple places at the same time.

Classical physics, based on the idea of continuous quantities and deterministic principles, struggled to account for these observed phenomena.

1.3 Quantum Mechanics and Quantum Field Theory

Now let's examine the two primary branches that compose Quantum Physics as a whole - they are:

- Quantum Mechanics and
- Quantum Field Theory.

Quantum mechanics explores the behavior of matter and energy at the smallest scales. It delves into the realm of subatomic particles, such as electrons and photons, and provides us with a deeper understanding of the fundamental nature of reality.

Its focus is addressed on the nature and behavior of the single particle and revolves around several key aspects, each contributing to its intricate tapestry of principles and theories. The wave-particle duality of matter and the probabilistic nature of quantum processes are the main topics of quantum mechanics.

Quantum field theory, on the other hand, combines the ideas of quantum mechanics with a field-based framework to explain particles and their interactions. Quantum Field Theory broadens the scope of Quantum Mechanics by examining the interactions between particles and the fields they generate, in addition to the behavior and characteristics of individual particles.

Quantum field theory tries to unify special relativity with quantum mechanics to create a cohesive framework for comprehending the fundamental forces of nature.

This field theory approach allows physicists to study particle interactions, such as those involved in particle colliders and high-energy experiments. It has been successful in predicting and explaining a wide range of phenomena, including the behavior of elementary particles, the creation and annihilation of particles, and the quantization of fields.

Quantum field theory has become an essential tool in modern theoretical physics and is a key component of the Standard Model, which describes the electromagnetic, weak, and strong nuclear forces.

I know, all of this sounds complex and ungraspable but we are here to unpack it. Chapter by chapter and piece after piece we'll understand the meaning of these words together.

We will start our journey with the first branch **Quantum Mechanics**.

SHEDDING LIGHT ON LIGHT

2.1 The Kickstart of Planck

We know that the classical physics of Newton and pals reigned supreme for a few centuries, but what was the specific thing that for the first time suggested there was more to the universe than we have previously thought?

This big shift was initiated in 1901 when Max Planck solved something regarding the electromagnetic waves called the *Ultraviolet Catastrophe*, a melodramatic name which we will explain shortly.

But first, we need to introduce in this extensive explanation the concept of **Blackbody** to better comprehend this Ultraviolet Catastrophe and how it all began.

What is a Blackbody?

A blackbody is a theoretical object that absorbs all incoming electromagnetic radiation, regardless of the wavelength or frequency, and does not reflect or transmit any of it.

If you want to know more about electromagnetic waves, see my publication: *"Basic Physics for Beginners - A Comprehensive Study Guide For The Self-Taught Scientist"*, there you have all you need to understand the basics of classical physics. In short, the electromagnetic spectrum encompasses all sorts of waves starting from the high frequencies such as gamma rays, X-rays, ultraviolet, and visible light, to the lower frequencies such as infrared, microwaves, and radio waves.

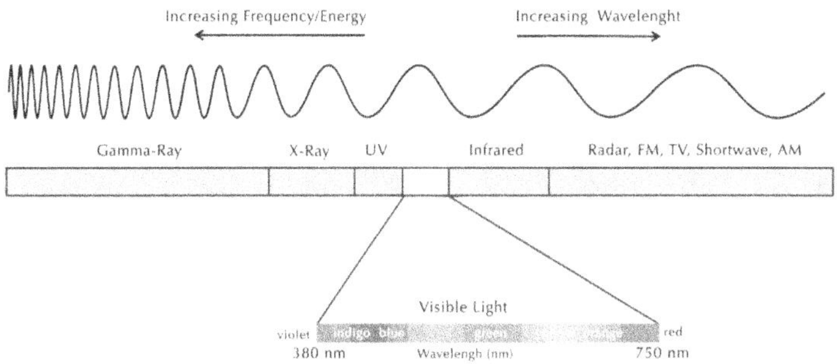

When we talk about visible light, we are referring though to the portion of the electromagnetic spectrum that is visible to the human eye. This portion ranges from approximately 400 to 700 nanometers in wavelength. Within this range, different colors are perceived by our eyes, including violet, indigo, blue, green, yellow, orange, and red.

So coming back to the blackbodies, these objects are called like that because they appear completely black when they are at a lower temperature than their surroundings. The concept of the blackbody was first introduced by the German physicist Gustav Kirchhoff in the 19th century.

When you shine a light on an object, some of it may be reflected, some of it may be transmitted and some of it may be absorbed.

Darker objects tend to absorb more of the light incident upon them and reflect less than lighter objects, which is why they appear dark in the first place but also why they tend to heat up more quickly - think about how black surfaces heat up in the summer compared with white surfaces.

If the temperature of a blackbody is less than the temperature of the environment in which it's placed then it will continue to absorb energy from the environment until it has reached the same temperature - at which state it is said to be in a thermal equilibrium.

If however the blackbody has a temperature that is higher than its surroundings, then it will radiate energy away until it has lowered its temperature sufficiently to reach thermal equilibrium.

Therefore a blackbody is not only a perfect absorber of radiation but also a perfect emitter of radiation.

While a true 100% blackbody doesn't exist in nature, it serves us as a useful theoretical model for understanding the behavior of radiation and how objects interact with it.

According to Max Planck, a perfect blackbody emits radiation at all wavelengths and frequencies, with the intensity of the emitted radiation depending on the temperature of the black body in question.

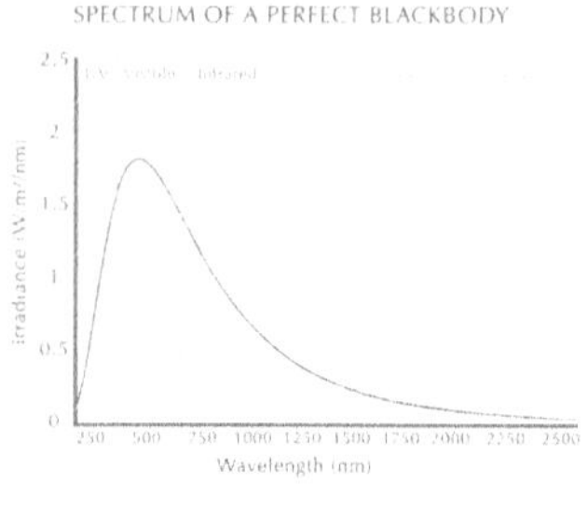

This relationship between temperature and radiation emission is known as blackbody radiation or thermal radiation.

Thermal radiation

Thermal radiation refers to the emission of electromagnetic waves from a heated surface. These waves carry energy in the form of heat and are commonly referred to as infrared radiation. Unlike conduction or convection, which require a medium for heat transfer, thermal radiation can occur in a vacuum as well.

When an object is heated, its atoms and molecules gain energy and start vibrating. As a result, they emit electromagnetic waves across a wide range of frequencies, including the infrared spectrum. This is what we perceive as thermal radiation.

Thermal radiation plays a crucial role in our everyday lives. For example, the warmth we feel from the Sun is a result of the thermal radiation emitted by it. Similarly, the heat we receive from a fire or a hot stove is also a form of thermal radiation.

Even if the perfect blackbody doesn't exist, the Sun, paradoxically, is the closest thing that exists in nature to a blackbody, that's due to the intrinsic characteristic of our star. The Sun has no solid surface, but any radiation that hits the star is scattered and absorbed until is completely lost in its nucleus. This makes the Sun very close to a blackbody, as far as absorption is considered.

Considering the emission of radiation instead, the Sun even if not perfect when compared to the model described by Planck, is a good radiator because emits strong thermal radiation, and if we overlap the spectrum of an ideal blackbody and that of the Sun in a graph, the curve would be similar in both bodies.

SPECTRUM OF A SOLAR RADIATION (EARTH)

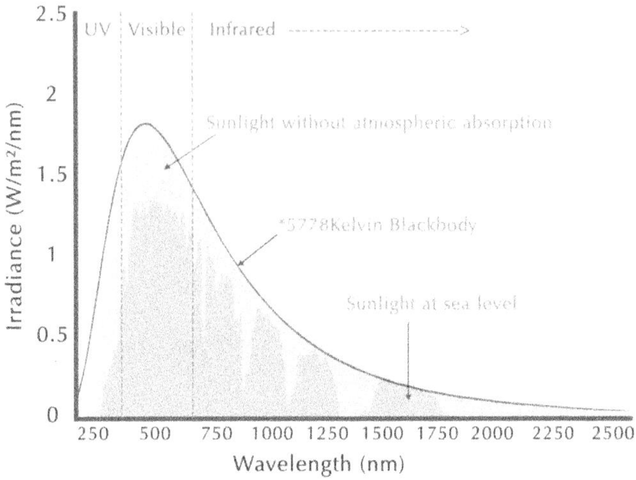

*The light emitted by the Sun (light grey) matches the blackbody spectrum for 5778K

If you take a look at the distribution of the wavelengths of light that we receive from the Sun, most of it as you can guess is from the visible spectrum, which is why our eyesight evolved to pick up this kind of wavelength, but we also receive light on either side, in the UV portion as well as infrared and beyond.

A hot rod of iron will also do this (emitting thermal radiation if stimulated), and this is the way we studied blackbodies at the time, noting that the distribution depends not at all on the material but only on temperature.

The more heat there is, the more the frequencies of the electromagnetic waves produced by the excited particles increase.

Resulting in a shift to the left part of the graph - as temperature increases we pass from lower frequencies (infrared) to the visible section of the spectrum.

This shifting explains why we start to see the iron change colors as it is heated up from black to red, to yellow, to white.

In other words, more heat = more movement of the particles = more shifting in the electromagnetic section of higher frequencies.

This dynamic will move into the visible spectrum at around 4000 Kelvin (3726°C) and above, this is why very hot objects appear to glow, like a hot oven, a light bulb filament, or the sun and other stars - because objects at these temperatures emit a lot of visible light, as opposed to something like me and you (a human body), which at around 310 Kelvin, emit essentially no visible light, which is why we can't see each other in the dark.

Color is defined by the average frequency. The reason the Sun appears yellow is due to the fact that its surface, which is 6,000 Kelvin, generates an incredible amount of photons within the green/yellow portion of the electromagnetic spectrum. Rigel, a blu supergiant star with a temperature of 12000 Kelvin, on the other hand, emits an abundance of ultraviolet radiation and high-frequency blue light.

The problem with the Blackbody Spectrum

The problem with the Blackbody Spectrum was that the classic electromagnetism could not account for it. Mathematical models attempting to produce these distributions were able to fit the data for longer wavelengths, but they did not predict that the intensity would dip down to the left for the UV portion of the spectrum as experimental data illustrated.

Mathematical logic predicted that the intensity would continue to increase as the temperature and frequency increase and become infinitely large as the wavelength approaches zero.

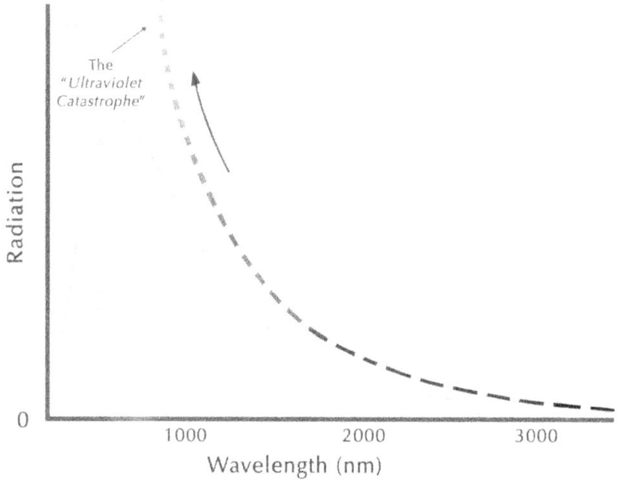

Of course, we know that can't be true, otherwise every time we use a toaster we would get blasted with UV radiation.

This contradiction was dubbed somewhat dramatically: *the Ultraviolet Catastrophe*. In a sense, it was catastrophic because it meant that something was fundamentally wrong with classical physics. In science, if a theory does not accurately align with observations of reality, it must be revised - and it was in this way that we realized that classical electromagnetism, as complete as it may seem, must have some kind of limitations in its ability to describe light and energy.

As we said, Planck solved this puzzle. The notorious physicist had an intuition and untangled the phenomenon by introducing a new brand concept called **Quantization**.

We know from classical physics that heat is just the transfer of kinetic energy from one place to another.

In the case of the hot rod of iron, for example, kinetic energy takes the form of atomic vibrations or oscillations. These spasmodic vibrations are what generate the luminosity we see in the blackbody spectrum.

Planck proposed that in order to avoid falling into the absurdity of an impossible and infinite accumulation of kinetic energy of a heated body - he presented the idea of circumscribed levels of energy that an electromagnetic wave can carry. That is a quantified minimum and maximum.

In other words, the energy of the electromagnetic waves they emit must be quantized.

This means that instead of having the ability to take on any value from a continuous series, they can only have particular and discrete values from a set of accepted values.

In this manner, he developed this expression for blackbody radiation: **E = nhf**

E = Energy

n = any integer (1,2,3..)

h = Planck's Constant (6.626 x 10^-34 J.s)

f = frequency of radiation

The **n** value is what results in quatization, as it can only be an integer, and not any fraction or decimal in between, meaning that the resulting energies will also comprise a set of allowed values, with anything in between being forbidden.

This application of the new quantization idea and the accompanying Planck's constant was only proposed for this specific practical purpose, but they allowed for the practical prediction of the true distribution of blackbody radiation at all wavelengths which meant that this constant is more than a mathematical fluke, but a clue as to the fundamental nature of reality.

And the fact that Planck's constant is so incredibly small, explains why the notion of quantization of energy had not cropped up before, because it shows that energy is "chunked" on such an incredibly small scale that the gradations between the allowed values are utterly minuscule. So as to appear non-existent to any measuring apparatus.

To give an idea of the scale of the dimension we're talking about - this is so small that if an atom was the size of the earth, a Planck length would be much smaller than even the head of a pin!

This explains why macroscopically speaking, we humans, and aliens with humanoid forms (such as Bohr and Einstein), see energy as continuous! Joking aside, on the fundamental level, it is indeed quantized, even though this conclusion was so strange that most scientists of the time, even Planck couldn't believe this sorcery of nature and that it had actual concrete physical meaning.

So, as we scale towards higher frequencies of vibration, the quantity of energy required to actually access the next quantized energy level (correspondent in the ultraviolet zone) is so colossal that the high-frequency vibrations cannot start vibrating - and this is what happens in our everyday life and this is why you're not blasted when using a toaster that has accumulated heat. To have any light emitted from the blackbody, the subatomic particles within the blackbody must be shaking with the specific frequency required to increase to the next level.

This means that eventually at a certain energy level, there's not enough thermal energy within the black body to induce the vibration at the frequency required for an increased energy. Therefore waves with very short wavelengths are not able to be generated and the blackbody spectrum shortens on the left side.

This was the first time the new brand concept of quantization had solved such a big puzzle in physics, but it wouldn't be the last.

Yeah but who was this guy?

MaxKarl Ernst Ludwig Planck was a German physicist born on April 23, 1858, in Kiel. He is best known for his groundbreaking work in quantum theory, which revolutionized the field of physics.

Planck began his academic journey by studying physics at the University of Munich and later at the University of Berlin. He completed his doctorate in 1879 and went on to pursue a career in academia.

Max Planck

In 1900, Planck made a significant discovery that transformed our understanding of energy. As we said he introduced the concept of quantization, which states that energy is not continuous but exists in discrete packets called quanta. This discovery laid the foundation for a new era in the field of physics.

Moreover, Planck's constant, denoted as '***h***', became a fundamental constant in quantum mechanics. It serves as a fundamental building block for calculating the energy of a photon or a particle, leading to the development of numerous technological advancements, such as lasers, transistors, and semiconductors.

Throughout his illustrious career, Max Planck not only made remarkable discoveries but also fostered a new era of scientific inquiry. His theories and contributions laid the groundwork for the development of quantum mechanics and revolutionized our understanding of the microscopic world. Planck's profound impact on physics continues to resonate to this day, inspiring generations of scientists and shaping the trajectory of scientific research.

Planck's innovative work on quantum theory earned him the Nobel Prize in Physics in 1918.

Impressive, but what is a Constant?

A constant is a value that remains unchanged and consistent throughout various experiments and observations. It serves as a building block for mathematical equations and theories, providing a framework for scientific understanding.

A constant in physics can be defined as a fixed numerical value that represents a physical quantity or property. These values are considered universal and are independent of any specific measurement or experimental setup. Constants are typically denoted by specific symbols or letters and are used to express the relationships between different variables in mathematical equations.

One of the most well-known constants in physics is the speed of light, denoted by the symbol '**c**'. It is a fundamental constant that represents the maximum speed at which information or energy can travel through space. The speed of light in a vacuum is approximately 299,792,458 meters per second (m/s).

Another important constant is the gravitational constant, denoted by the symbol '***G***'. It represents the strength of the gravitational force between two objects and is a key component in Newton's law of universal gravitation. The gravitational constant is approximately equal to 6.67430 × 10^-11 cubic meters per kilogram per second squared (m^3/kg/s^2), and it determines the magnitude of gravitational interactions between celestial bodies.

Constants in physics can differ in their nature and application. Some constants, like the aforementioned Planck's constant (***h***), are fundamental and have a broad range of applications across different branches of physics. The Planck constant represents the quantum of action and is used to describe the behavior of particles on the atomic and subatomic scale.

Ok, but how did Planck came out with this constant?

In classical physics and maths literally everything can be infinitely divided, there's no smallest of anything. Before Planck's arrival on the stage, the calculations, in theory, allowed the particles to vibrate with any amount of energy, all the way down to infinitesimally tiny wiggles.

If for a moment we tried to distribute heat energy to equipartitionates across possible energy states, way too much energy was packed into the countless very tiny energy states at high frequencies. There's nothing wrong with it mathematically, but the result is absurd in real life. In infinitely dividing the smallest remaining energy states you finish never finding an end to them.

First, we must say that Planck resolved the catastrophe almost by accident while searching for a new mathematical approach to deriving the blackbody spectrum - he needed some sort of math gimmick to count the supposedly infinite energy states. In what he described as a moment of desperation, he tried something pretty counterintuitive.

He decided that those particles could only vibrate with energies that were a multiple of some minimum energy. As we said, he quantized the energy states. He set this minimum energy to be the frequency of a particle's vibration times a very, very small number a number that at that time was yet to be measured.

That number became the Planck constant.

It was akin to haphazardly pressing a keyboard key when your computer stalls, yet the maneuver proved effective. Why did it enable Planck to solve the UV disaster by equipartitioning energy? Because it restricted the amount of energy that those vibrations at high frequencies might contain.

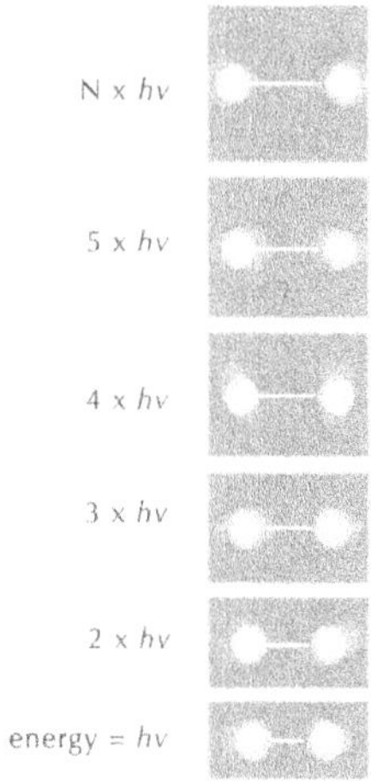

Now, Planck initially believed that these quantized energy levels were only mathematical tricks. He expected that his new constant would come out to be zero, indicating the absence of quantization, minimal energy, and the interval between energy states.

If such had been the case, the Planck constant would not have been included in the final equation. On the contrary, the Planck blackbody law is deeply embedded with this recently discovered constant. Quantization of energy is real!

Now, the actual value of the Planck constant still had to be measured but given a mathematical form for the blackbody spectrum, it was possible to do this just by varying the constant until the Planck law matched the observed spectrum.

But in the end, it took Einstein to fully understand the physics behind Planck's strange quantized vibrations. Einstein realized that it's actually light that is quantized. Those little crazy particles do have quantized energies, but it is because they can gain or lose energy by absorbing or emitting one particle of light at a time and that light comes in indivisible energy parcels.

2.2 The Photoelectric Effect

Now let's see the photoelectric effect, another important piece in this immense mosaic of quantum mechanics. This discovery together with the double-slit experiment are those that expose the true nature of light and its duality.

As we've mentioned before, it was Einstein who implemented and delved more into the work initiated by Planck arriving at some astonishing discoveries and deductions.

Einstein proposed that the energy carried by a single photon is directly proportional to its frequency. This means that higher-frequency photons carry more energy than lower-frequency ones.

But what is a photon? It's a kind of electron but smaller?

The photon is the fundamental particle of light and electromagnetic radiation in general, is the smallest and indivisible piece of any electromagnetic manifestation.

In the quantum field, it is defined as a discrete bundle or quantum of electromagnetic energy. Photons are considered massless particles and travel at the speed of light in a vacuum.

The concept of photons was first introduced, guess who? Correct always him, by Albert Einstein in 1905, as part of his explanation of the photoelectric effect. Einstein proposed that light consists of discrete packets of energy, precisely called photons.

They can interact with matter, get absorbed or emitted by atoms, and carry specific amounts of energy.

On the other hand electrons manifest both particle and wave-like properties, they are considered fundamental particles that carry a negative electric charge. They are a part of every atom and are responsible for the flow of electric current.

Photons are packets of energy while electrons are a part of the matter itself, they have mass, although smaller compared to other subatomic particles like protons and neutrons.

Here are the key differences to avoid confusion:

- Photons regard the electromagnetic, electrons regard the matter;

- Photons are massless particles, while electrons have a small mass;

- Photons travel at the speed of light, while electrons have varying speeds depending on their energy levels;

- Photons have no charge, while electrons carry a negative electric charge;

- Photons are emitted or absorbed by charged particles during transitions, while electrons are involved in chemical reactions and electric currents.

So, when light interacts with matter, it can transfer energy to the electrons within the surface of the material. The energy carried by photons in the light can be absorbed by electrons, causing them to be excited to a higher energy level or even completely detached from the material, resulting in the emission of electrons (being they the part of atoms the most susceptible to external forces and motions).

The photoelectric effect is particularly interesting due to its counterintuitive nature. According to classical physics, it was expected that the intensity of light would determine the energy of the emitted electrons. However, experiments conducted by Einstein and others in the early 20th century revealed that the energy of the emitted electrons actually **depends on the frequency** of the incident light rather than its intensity.

Let me explain it better:

In Einstein's time and centuries before him, light was thought of as clearly having wave-like properties (diffraction, interference, and so on). However, every once in a while it seemed to behave in a very unwave-like fashion.

Some experimenters found that if you shine a light on a metal surface just in the right way instead of illuminating or warming it, a bit of the light would hit it like a tiny cannonball knocking an electron out of the surface.

This effect was first demonstrated in the 1800s but at the time no one could explain why this happened - if light were just a wave of energy, shining a light on the surface should either do nothing or it should gradually transfer enough energy to push electrons around. Higher-intensity light should of course dislodge more electrons, but the experiments with the light on the metal showed that it didn't matter how long the light was shining or even how intense it was. <u>All that mattered was its color.</u>

A light with a low frequency (like infrared or visible light) would do nothing but if you gradually increase the frequency into the ultraviolet, suddenly electrons would start pinging off the metal like they were shot with tiny light pellets.

That's not how a wave is supposed to work!

And it was at this moment that Einstein came in and said "Hey! Can't you see? Light isn't a pure wave, it is made of tiny particles too!" This little particle (photon), carries an amount of energy determined by the light's frequency or color where it belongs.

A photon of visible light has too little energy to knock an electron out of the metal, even if you shine a lot of visible light photons on the metal, each photon will just shift an electron to a higher energy level after which it will fall back down.

Because the energy levels of electrons only exist in discrete steps, you need a photon of just the right energy to remove the electron completely and for that, you need to dial up the light's frequency into the ultraviolet so each photon can pack enough of a punch.

One significant application of the photoelectric effect is in solar energy systems. Photovoltaic cells, commonly known as solar cells, are composed of materials that exhibit the photoelectric effect, such as silicon. When sunlight falls on the solar cell, the absorbed photons of light create an electric current by releasing electrons from the material, which can then be harnessed for power generation.

Another area where the photoelectric effect plays a crucial role is in digital imaging technology. Digital cameras and image sensors rely on the conversion of light into electrical signals through the photoelectric effect. When light enters the camera lens, it falls onto an image sensor that contains millions of tiny light-sensitive elements called pixels.

They use a material called semiconductors in which even low-frequency photons can boost the electrons, so all the light that hits the sensor can create an electrical current. Filters, lenses, and materials optimized for different frequencies work together to record the color of the light - once the image is recorded by the sensors, the signals can be re-encoded into an image for your screen.

2.3 The Double-Slit Experiment

The second most well-known and fascinating experiment in the history of quantum mechanics is the *Double-Slit* experiment. This experiment combined with the *Photoelectric Effect*, is the backbone of theories about the nature of light.

So as we saw, there has been a strong focus in the previous two centuries on understanding the mysteries of light, its nature, and its mutable behavior, but it must be said that this debate had been ongoing for centuries before Einstein, with prominent physicists like Isaac Newton (advocating for the particle theory), while others, like Christiaan Huygens, for the wave theory.

However, it was not until the early 19th century that a significant discovery occurred. Thomas Young, an English physician, and physicist, conducted his famous double-slit experiment with the aim of putting an end to this very long diatribe.

Thomas Young

Pay attention to this experiment, dear reader, because it was this experiment that opened the Pandora's box and led the entire scientific world to incredulity. It's so true that... is still called today *"The Central Mystery of Quantum Mechanics"*.

The Setup of the Experiment

In this experiment, a beam of monochromatic light or a stream of particles, such as electrons or photons, is directed toward a barrier that contains two closely spaced micro-slits. When the beam passes through the slits, it creates an interference pattern on a photosensitive screen placed behind the barrier.

In the experiment, it was immediately observed that the monochromatic light, (or light with a single wavelength), behaves nearly exactly like a classical wave when it is shone through the two slits in the barrier.

Most of the wave is blocked, but ripples pass through the slits and when a new ripple starts to overlap with the previous one, they produce a mesmerizing pattern that is called ***interference pattern***.

What we see is that as the waves spread out - when the crest of one wave hits the crest of another, it becomes stronger and that sum is called constructive interference, where instead in the same propagation a crest hits a trough, it will cancel itself and it's called destructive interference. This alternation of constructive and destructive interference once hits the screen creates the pattern observed. But this should be no mystery to us, this is not quantum mechanics, this is just the classic property of waves. It happens with water waves too.

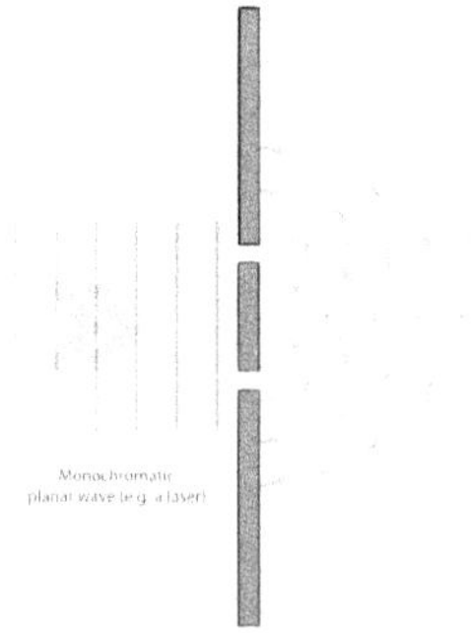

So what we see produced on the screen are a kind of bands of light and dark stripes which coincide respectively in regions of constructive and destructive interference.

Light waves passing through the slits

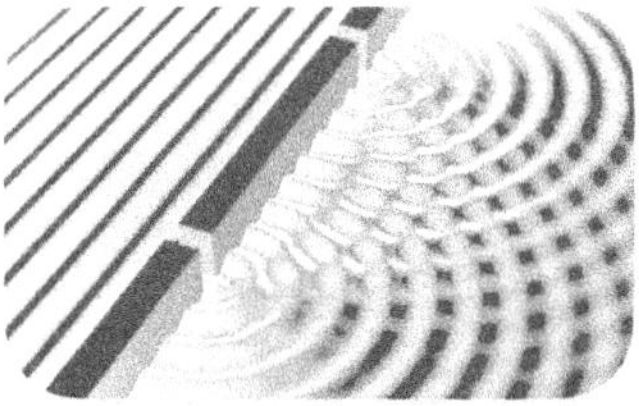

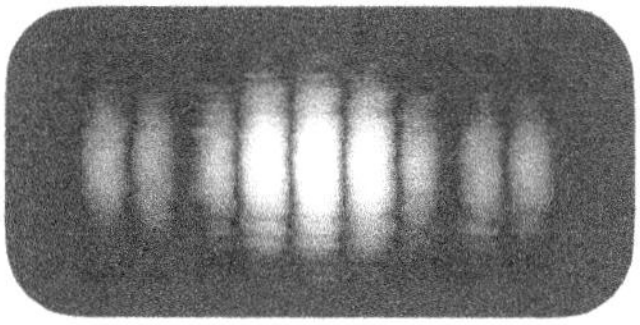

Therefore the most immediate response following this evidence is that light is simply **a wave**. Furthermore, we now know that light is a wave in the electromagnetic field thanks to Maxwell's work a century later, so it makes perfect sense that it should produce an interference pattern, right?

Ladies and Gentlemen of the Jury, the case is closed.

But wait!

Let's not get carried away by easy enthusiasm. What if we lower the intensity of light, and keep lowering it until our light source emits only one photon at a time?

Surely shooting only one photon at a time shouldn't create a wave pattern, ...right?

When we shoot one photon at a time, it forms a dot on the screen. But if you fire enough of them, constantly, thousands or millions of

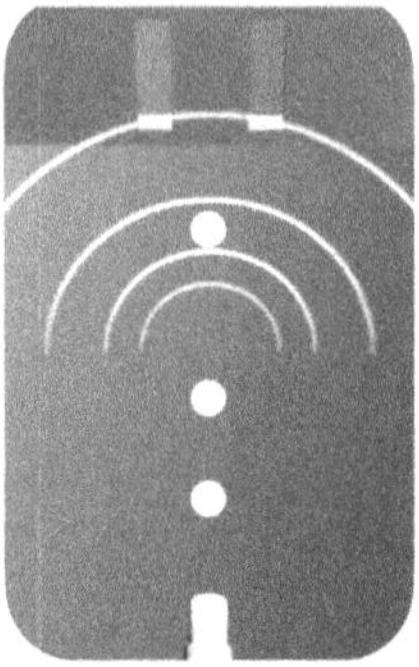

photons, what you find is that collectively, the photons make emerge the same interference pattern.

The pattern emerges in the distribution of final position of many completely unrelated photons, but how can that be?

Every single photon doesn't know where other photons have landed or will land; however, before it reaches the screen, it knows which areas have the best chance of landing and which ones have the worst chance.

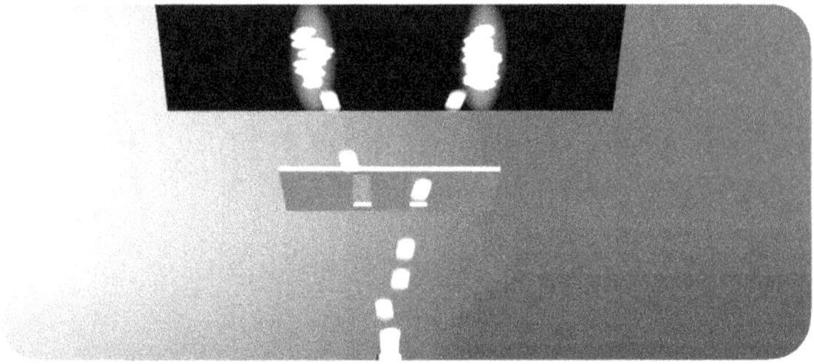

It does this by analyzing the interference pattern of a pure wave that equally passed through both slits and choosing its landing spot accordingly.

Therefore, when fired individually, <u>each photon either interacts with itself or conveys its role in the interference pattern</u>.
What if, in our quest to understand the slit experiment, we installed a detector to discern whether the photon passes through slit 1 or slit 2, and if so, why does it deflect?

Well, here comes the most **SHOCKING** result of quantum mechanics: when we try to find out which path the photon took, the **interference pattern ceases to manifest itself**!
Come on! Seriously?!
The photon now appears to behave like a <u>particle</u>, instead of a wave. On the screen, we only see two strips corresponding to the two slits on the barrier. So what's GOING ON here!?

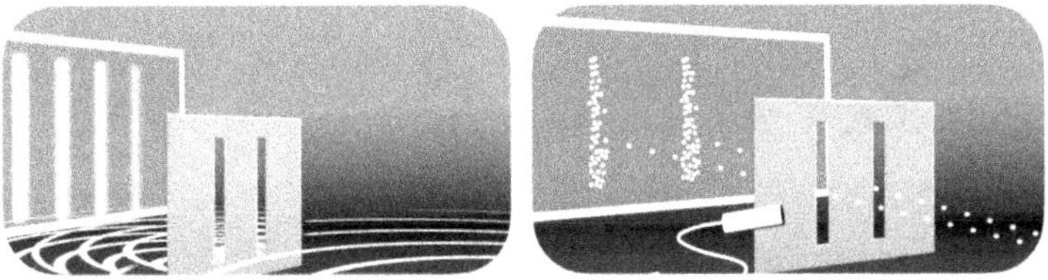

All of this started with the aim of arriving at a definitive clarification but we found ourselves in a situation of further confusion!! And as said all of this was before the advent of the Einstein era.

Just let me anticipate you a couple of concepts.

Speaking in terms of the most widely held view of quantum mechanics, the ***Copenhagen interpretation***, around 1927, two particle nerds at the University of Copenhagen named **Niels Bohr** and **Werner Heisenberg** came up with this interpretation, which essentially states that these so-called particles are really more like "clouds" in that they exist in multiple states simultaneously.

To be more precise, they are waves of probabilities.

What on earth...

Yes! There exists a probability where is much more probable that the photon behaves much more like a particle and one where is more probable that it behaves more like a wave.

When a probability wave "**collapses**" (this is the term used) into a discrete and distinguishable particle—that is, when it is measured —the waves become distinct, like particles. The probability wave of any particle collapses at any moment when its which-path information is examined, according to physicists.

Now hold your skull in because your brain is going to explode with the following summary sentence:

Particles can only continue to be probability waves if their paths are completely secret, meaning no records exist!

What was observed was that the particles become discrete/ distinguishable and stop behaving like wave-light **if** any form of measurement is done. Even electrons, atoms, and molecules with up to 800 atoms have all exhibited the same behavior, which is crazy. If their which-path information in the experiment is measured, the same interference pattern arises and the same collapse occurs.

A measurement always collapses the wave.

Thus, how can we define a measurement? It's legitimate to inquire, right?

A measurement is a formation of any physical record of which path the particle takes. It's not about the conscious observer, it's about a record (ANY record) made in the universe even if no one is actively looking.

As said, it's like the path has to be taken in absolute secrecy, the particle has to be informationally isolated. The interference (the wave-like manifestation) can only occur if it's **impossible**, even in principle, to find out which path the particle took. Bizzare, right? And this is only the beginning.

THE ATOM AND ITS PARTICLE DYNAMICS

Within the atomic realm, quantum mechanics provides us with a detailed understanding of the energy levels and electronic structure of atoms. The quantization of energy levels allows us to explain why electrons occupy specific orbits around the atomic nucleus.

This discrete energy structure is in stark contrast to the continuous energy levels predicted by classical physics. Knowing the configuration of the atom will help us to comprehend better more concepts later. But before going any further, let's dive a little bit into the history of the atom, its representation throughout time, and what is today's model, shall we?

3.1 The Hellenic one

In the history of science, various models have been proposed to describe the structure and characteristics of atoms. These models have evolved over time as our understanding of atomic structure has advanced.

One of the earliest models of atoms was proposed by the ancient Greek philosopher Democritus in the 5th century BC as mentioned in Chapter 1. He suggested that atoms were indivisible and indestructible particles that made up all matter. The word "atom" can be traced back to the Greek word "*atomos*," which precisely means "indivisible" or "uncuttable."

In his conviction:

- all matter consists of atoms; these are too small to be seen;
- there is empty space between all atoms;
- atoms are solid but have no set internal structure;
- every substance is made of different atoms, with their own weight, size, and shape.

Despite the Democritus' model was purely speculative and lacked any experimental evidence, it was certainly avant-garde for its time.

3.2 The Thomson's pudding

J.J. Thomson

Centuries passed by without any major serious investigation regarding these little blocks at the base of the matter, being recorded. This was until the late 19th century, when the English physicist Joseph John Thomson, famous for discovering the electron, proposed the "plum pudding" model of the atom. According to this model, atoms were composed of a positively charged "pudding"

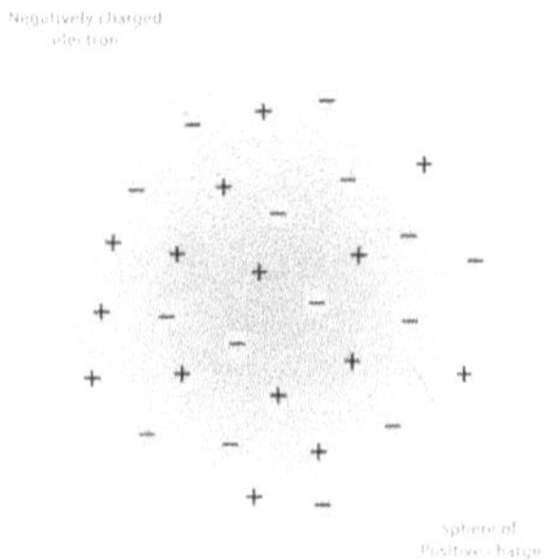

Thomson's plum pudding atomic model

with negatively charged particles embedded in it, much like plums in a pudding.

We are in 1897 and J.J. Thomson's experiments involved the study of the behavior of cathode rays in a vacuum tube, which led Thomson to propose the existence of a fundamental particle: the electron precisely.

Thomson's experiments involved the use of a cathode ray tube, a sealed glass tube with electrodes at each end. By applying high voltage across the electrodes, a beam of cathode rays was produced.

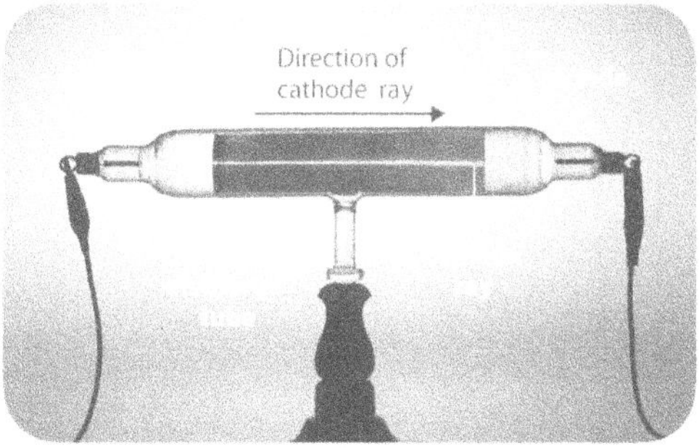

Through careful observation and experimentation, Thomson made several significant discoveries.

One of the key findings from Thomson's experiments was the observation that cathode rays were deflected by both electric and magnetic fields. When a magnet was brought close to the beam in the tube, it deflected (towards the magnet when it was brought close to the positive pole, away from the magnet when brought close to the negative pole).

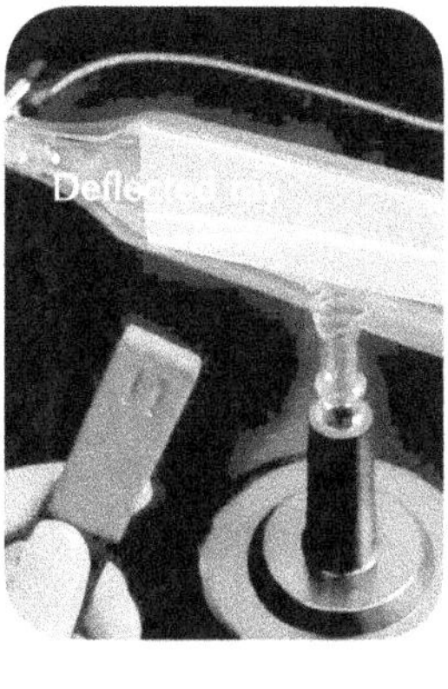

This deflection indicated the presence of something negatively charged within the cathode ray, he thought most probably particles. Thomson hypothesized that these particles were fundamental components of atoms.

He also said: “We may use any kind of substance we please for the electrode and fill the tube with gas of any kind and yet get the same velocity in which these particles travel.”

This was revolutionary! Thomson concluded that everything was filled with these tiny little things that he called “corpuscles” but we now call electrons - and the corpuscles in one item are exactly the same as the corpuscles in another.

Thomson concluded that atoms were **not indivisible**, but rather composed of smaller, negatively charged particles. This led to the first level of complexity in the development of the atomic model. According to this "plum pudding model" or "raisin bun model", electrons were embedded within a positively charged sphere, much like raisins embedded in a bun.

3.3 Rutherford: matter is mostly empty space

Thomson’s model couldn’t explain the results of several other experiments with really bizarre results though, one of these experiments was done by the New Zealand physicist Ernest Rutherford.

This experiment showed that most of the mass of an atom is concentrated in a tiny, positively charged nucleus at the center of the structure, with electrons orbiting around it. This led to the development of the "*planetary*" model. Rutherford’s experiment got popular by the name of the *“Alpha particle scattering experiment”*.

E. Rutherford

But first, let’s see what *Alpha* particles are.

Cool! And a Radioactive Substance? What is that?

To put it simply, a radioactive substance refers to any material that contains unstable atomic nuclei, leading to the emission of ionizing radiation. These substances can be found naturally in the environment or created artificially through nuclear reactions.

Radioactivity occurs when an atom's nucleus is unstable, meaning it contains an excess of either protons or neutrons. In order to achieve a more stable state, the atom releases energy in the form of radiation. This emitted radiation can take various forms, including alpha particles, and gamma rays.

One of the most well-known radioactive substances is uranium. Uranium is a naturally occurring element that undergoes radioactive decay, transforming into other elements over time. This process releases radiation, which can have both beneficial and harmful effects depending on its usage and exposure.

Alpha particles are generally emitted by radioactive substances like helium ions (characterized by having a double positive charge).

So, an **ion** is an atom or molecule that has **gained** or **lost** one or more electrons, resulting in a positive or negative charge as a whole. In this case, *Alpha particles* are particles positively charged.

But let's go back to our Rutherford experiment. The setup of the experiment consisted of:

- A lead box in which is placed the source of the ions positively charged;
- A hole in one of the box walls;
- A thin gold foil (which will be at the center of our observation);
- A fluorescent curved screen around the foil.

The alpha particles emitted from the radioactive source pass through the hole of the box and enter into the center of our observation - the thin streams of the particles then bombard the gold foil and get scattered in different directions.

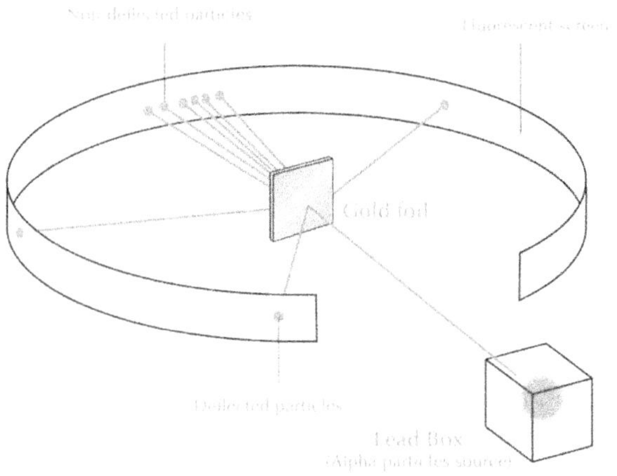

A screen coated with zinc sulfide to make it fluorescent was used as a counter to detect alpha particles. Each alpha particle struck the fluorescent screen, causing a burst of light known as scintillation, which was seen through a viewing microscope attached to the rear of the screen. The screen itself was moveable, allowing Rutherford and his colleagues to see whether the gold foil was deflecting any alpha particles.

Such scattering patterns were shown by the detector. And here happens something really peculiar and surprising for the expectations at that time.

It was expected from Thomson's model (which is basically a sphere positively charged with negative charges randomly placed within it), that it should show a large number of deflections but something else was observed actually.

The detector showed that most of the alpha particles passed through the gold foil as such, without showing any deflections. Only a few of them showed deflection in different directions, a very few of them, to be precise one in 11,000 rebounded back.

As most of the particles passed undeflected, Rutherford concluded that **most of the space inside of the gold atom was empty**, and a few particles were deflected from their path by small angles indicating that there must be a positively charged center, which was repelling alpha positive particles, as these particles were few in number, this led to the conclusion that a positively charged center occupies a tiny space.

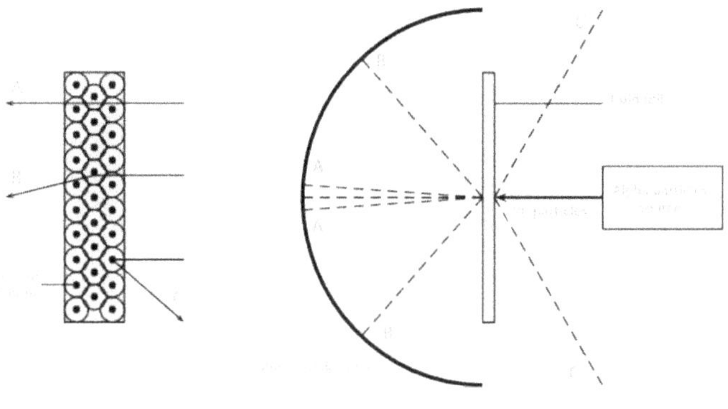

The small number of these particles further led Rutherford to conclude that the positively charged center occupies a miniature volume. All of the atom's positive charge was concentrated in a tiny volume in its core, as seen by the limited number of particles that bounced back. Rutherford used the term "*nucleus*" to describe the positive core of an atom. Based on his calculations and observations of the particle proportions (undeflected, deflected, and rebounded), Rutherford determined that the nucleus contained the vast majority of an atom's mass. In light of these findings, Rutherford proposed a model of the atom that would come to be known as the *nuclear* or *planetary* model.

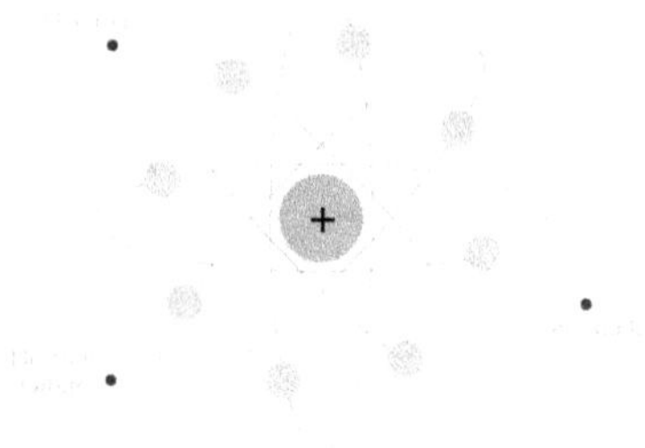

Rutherford's planetary atomic model

In a nutshell, according to Rutherford's nuclear model, there's a positive charge at the center of the atom called the nucleus and all of the mass of the atom is due to this nucleus, it also suggested that the electrons of the atom revolve around its nucleus in circular paths called orbits.

Despite its significant contributions in terms of popularizing how an atom was structured, the Rutherford atomic model had a few limitations. It failed to explain the stability of atoms, as according to classical electromagnetic theory, the revolving electrons should continuously lose energy and eventually fall into the nucleus and hence the atom should collapse (which doesn't happen, nature doesn't implode constantly). Additionally, this model could not account for the existence of atomic energy levels and the spectral lines observed in atomic spectra, later studied in detail by Bohr.

N. Bohr

3.4 Bohr's energy levels

In 1913, the Danish physicist Niels Bohr introduced an upgraded model of the atom, which expanded upon Rutherford's planetary model making it more stable. Bohr postulated that:

1. Only certain special orbits called also discrete shells are allowed inside the atom;

2. While revolving in these discrete shells, the electrons do not radiate energy;

3. There exists a defined energy level within every orbit or shell. This is why orbits are called stationary orbits or stationary shells - orbits or shells are also known as *energy levels*.

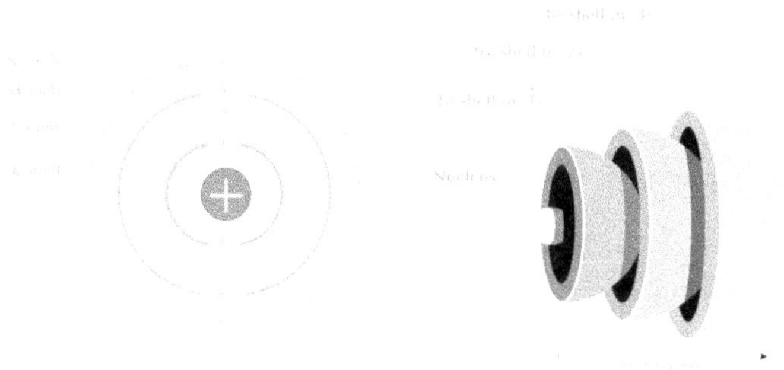

Beginning from the closest one
to the nucleus, shells are represented by the letters, "K", "L", "M", "N", or the numbers "1", "2", "3", "4" and so on depending on the number of the shells in the atom.

With this model he successfully provided a framework to understand how electrons transition between different energy levels, emitting or absorbing specific wavelengths of light in the process.

Let's take for example the atom of hydrogen (which the Bohr model is built on), we have these energy levels on which an electron can be - the energy level one (n=1), energy level two (n=2), energy level three (n=3), but it can never be found in the middle because it's quantized as Planck suggested, it has to be in one of these levels mentioned or another superior level. And so how does the electron move between levels?

Well if it absorbs energy from a photon e.g., electromagnetic radiation - it will suddenly raise its energy to the next level, only to fall immediately afterwards to its original level. During the "fall" the electron will in turn produce other photons.

This theorization helped us comprehend the colors we see in the spectrum; each photon has its own unique energy packet, and these packets correlate to different wavelengths, therefore the colors we see vary.

Let's see this concept better.

We need to introduce something really essential to understand better this atom model, we need to have at least a basic grasp of what is referred to as *The Series of Spectral Lines.*

Spectral lines refer to the specific wavelengths of light emitted or absorbed by atoms when their electrons transition between different energy levels. In the case of hydrogen, these transitions give rise to distinct patterns of spectral lines.

One of the most well-known series in hydrogen spectra is the Balmer series. It consists of a set of spectral lines that appear in the visible region of the electromagnetic spectrum. These lines are characterized by their unique wavelengths, which can be mathematically described by a formula, formulated by the Swiss mathematician Jacob Balmer.

If for example, an atom of hydrogen is absorbing energy and an electron starts to jump let's say from the level n=2 of the orbits to higher energy levels then this electron can drop back down to its

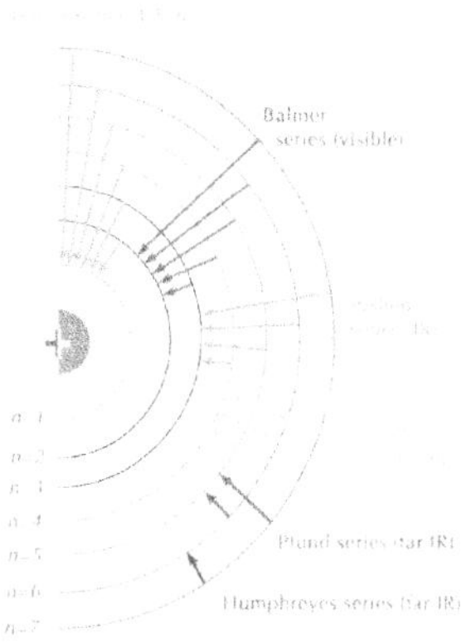

starter level (n=2) and release a very specific and discrete amounts of energy.

Let's say this electron descends from level n=3 to n=2, the ***Balmer series*** observation indicates that this descent would manifest a photon with a lengthwave that would correspond to the red part of the spectrum.

If it jumps down instead from n=4 to n=2, the wavelength would therefore be a little smaller, and as a result, it would correspond to orange (due to its higher frequency) - if it jumps down from n=5 to n=2, we would get yellow and so forth. Now we don't only have these Balmer series (that concern only the visible light), there are also hydrogen emission spectra and absorption spectra for the ultraviolet and infrared series.

The ultraviolet series is referred to as the Lyman series whereas the infrared series are called the Paschen series - and what it means is that instead of the value n=2 to be the first starter value, for the Lyman series, you get the n value being 1, for the Paschen series you get them dumping down to 3.

This model holds significant importance in the field of atomic history as **it was the first to propose the concept of quantum mechanics**, incorporating discrete energy values.

What we see is the Lyman series corresponds to lines on the ultraviolet section and the Paschen would correspond to the infrared section.

So in using and understanding these spectrum lines, Bohr was able to verify its postulates, in terms of the model of the atom.

In summary, now we have a model of the atom that is far more robust than Rutherford's initial model - to this planetary configuration was implemented a further dynamic, which is, as explained, the jumping between orbits, almost like a ladder, the electron must be in a specific ladder step to exist.

However, Bohr's model does have some problems too and we're going to look a them.

Firstly, the Bohr model fails to explain the underlying mechanisms that govern the stability of electrons in their orbits. According to classical mechanics, a stimulated electron should emit electromagnetic radiation and eventually spiral into the nucleus (since the two particles that make up the atom are naturally oppositely charged). This contradicts the observed stability of atoms in general and their ability to maintain their energy levels over time.

Secondly, the Bohr model does not account for the wave-like nature of electrons, he explains the photons emitted with their specific wavelength but treats the electrons as particles confined to specific orbits, without considering their **wave-particle duality** too. This limitation becomes more apparent when studying the behavior of electrons in complex systems or at high energies.

Another limitation of the Bohr model is its inability to explain the fine structure of spectral lines. The model predicts the existence of discrete energy levels and corresponding emission or absorption lines, but it does not account for the slight variations observed within these lines. This discrepancy is attributed to the effects of electron spin and relativistic corrections, which are not considered in the original model.

Furthermore, the Bohr model is limited to hydrogen-like systems, where only one electron is present. It cannot accurately describe multi-electron atoms or molecules, which require more sophisticated quantum mechanical approaches.

3.5 The "Swiss" Chadwick Particle

In history, due to the complexity of the matter, scientists had to do a whole variety of experiments in order to discover any possible piece that may compose the atom, one by one including their behavior. And I don't know if you noticed but one piece is still missing from our head count.

The **Neutron**. The neutron was the last of the subatomic particles to be discovered, in the first two decades of 1900. As we've seen, the model consisted of a nucleus in the center of the atom positively charged, composed of protons but no neutral particles had been mentioned yet.

Now, there were a few clues that led some of the scientists to ponder that something like neutral pieces might existed in the nucleus of the atom.

First of all, it was observed that all these protons in the center of the structure had no instability of any sort. How could that be? Protons are positively charged and like-things, as magnetism dictates, repel! If we got two positively charged particles they will want to push away from each other.

So it was thought, that maybe there's something else that helps keep these positively charged particles together, keeps them from just pushing away and repelling each other - and it turned out that it is exactly what neutrons do.

This Swiss particle as I call it, is neither positive nor negative and its role is essential in order to maintain a balanced and stable structure.

Another clue popped up when scientists compared the masses of Hydrogen(H) and Helium(He), finding interesting details. It was already known that each hydrogen atom had one proton and that each helium atom had two protons - it was decided by convention that one proton should weigh 1 a.m.u. (atomic mass unit).

So they thought that the two protons of two atoms of hydrogen, like in this example, would weigh 2 amus and the two protons of the single atom of helium should weigh 2 amus too, so 2 amus of both sides of the scale.

But it turned out that one **He** atom was still heavier than 2 atoms of **H** and this didn't make sense because they knew that there was an equal number of protons on both sides of the scale. It took 4 **H** atoms to balance out 1 **He** atom. Of course for us now the answer is very intuitive, but for those researchers at the time in the 1910s-1920s, this was a real head-scratch. The answer is that helium doesn't just have protons in its nucleus, helium atoms have 2 protons and 2 neutrons, so four particles in total, this explains why 1

He atom balances out the mass of 4 protons in the 4 **H** atoms.

The same thing happened with other elements, like Lithium with an atomic number 3 (which is the number of protons) having a total mass of 7 amus, Beryllium (atomic number 4) 9 amus, Boron (atomic number 5) 11 amus and so forth. The atomic number never matched with the total mass - it became clear that something else in the nucleus that has mass makes the atom weigh more than you might expect.

Around this time, two German scientists, Walther Bothe, and Herbert Becker, found that when they bombarded beryllium with alpha particles (which are basically helium nuclei), a strange type of radiation was emitted. This radiation was highly penetrating and couldn't be explained by the known particles at the time. They thought it might be a new kind of gamma radiation, which is high-energy photons. But it was a scientist by the name of James Chadwick who came along in 1932 and figured out exactly what neutrons were. He decided to investigate this mystery further. Here's how he conducted his experiment: He took a surface of Beryllium, and shot alpha particles from a polonium source, kind of in the same way Rutherford shot alpha particles at the gold foil.

J. Chadwick

Chadwick shot alpha particles at the surface and surprisingly observed the emission of a unique form of radiation that was not influenced by electric or magnetic fields. This radiation was neither negatively nor positively charged and as said, this radiation possessed a high penetrating power and was capable of penetrating through lead that was several centimeters thick.

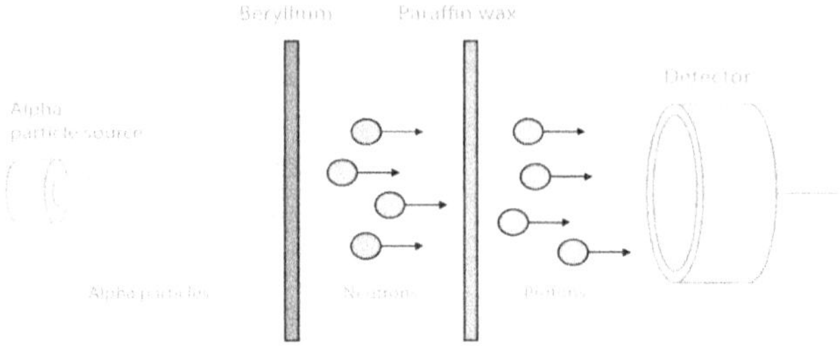

After the beryllium, Chadwick placed different materials, including paraffin wax, in the path of the resulting radiation to see what would happen.

When the mysterious radiation hit the paraffin wax, something interesting happened: protons were knocked out of the wax.

Now, this was unexpected because it is well-known among physicists that gamma rays wouldn't typically have enough force and momentum to knock protons out of atomic nuclei like that.

Chadwick realized that if this radiation were actually made up of uncharged particles with a similar mass to protons, they could easily knock protons out of the paraffin wax. These neutral particles wouldn't be deflected by electric or magnetic fields, which would explain why they were previously mistaken for gamma rays.

To confirm his theory, Chadwick repeated the experiment with different materials and carefully measured the energies and the angles at which the protons were ejected. The results matched the idea that these neutral particles were about the same mass as protons but had no electric charge. Through a combination of clever experimentation and careful thinking, Chadwick unveiled the neutron, fundamentally changing our understanding of atomic structure. It was in this way we had the identification of a new particle placed in the nucleus, and for this groundbreaking work, Chadwick was awarded the Nobel Prize in Physics in 1935.

THE REALM OF SCHRÖDINGER

This chapter is a continuation of the previous one, we will see how the evolution of the atomic model continues up to this day. We dedicate a special chapter to Schrödinger because his contribution in this field was nothing short of impressive - he officially incorporated the wave-particle duality in the atom, introduced the concept of electron orbitals, the concept of superposition and developed the famous equation that carries his name - furthermore, he helped to create a framework for understanding phenomena like electron tunneling and quantum entanglement and yes of course, his cats!

But let's proceed in order.

Before delving into Schrödinger's scientific achievements, it is essential to briefly explore his early life and education. Born in Vienna, Austria, Schrödinger displayed an aptitude for physics and mathematics from a young age. He pursued his higher education at the University of Vienna, where he studied physics under the guidance of notable scientists such as Franz Exner and Friedrich Hasenöhrl.

Schrödinger's career gained momentum when he was appointed as a professor at the University of Zurich in 1921. It was during his time in Zurich that he formulated the wave equation. One of the most intriguing aspects of Schrödinger's work is his involvement in the development of the

Copenhagen interpretation of quantum mechanics.

This interpretation, proposed by Niels Bohr and Werner Heisenberg, sought to explain the behavior of particles at the atomic level. Schrödinger's wave equation provided a mathematical framework for the Copenhagen interpretation, and his efforts played a crucial role in establishing it as one of the prevailing interpretations of quantum mechanics.

In addition to his contributions to physics, Schrödinger also had a keen interest in philosophy and the intersection of science and spirituality. He explored the concept of consciousness and its connection to quantum mechanics.

Schrödinger's work earned him numerous accolades, including the Nobel Prize in Physics in 1933, which he shared with Paul Dirac. Despite his success, Schrödinger faced challenges during his career, including political unrest in Europe and being forced to flee from the Nazi regime due to his opposition to their ideologies.

4.1 The Superpositioned Cat

Let's start with the most widely-known example in this branch of quantum physics. The Schrödinger's cat.

This illustration has become so far-famed that it has almost achieved the rank of a meme in the field. This thought experiment though, is not a real experiment, so please DO NOT try this at home, at least not with cats or any other living creature. A politician maybe..

Joking aside, at this point in quantum mechanics, particles such as electrons, photons, and even entire atoms have been hypothesized could exist in a state of superposition.

But what does it mean? It means that they can exist in multiple states at the same time. The idea behind Schrödinger's cat relates to a very real observed phenomenon, that is, in the quantum world sometimes, a particle appears to be doing two contradictory things at the same time.

Unlike classical physics, where particles have well-defined properties, quantum particles can be in a combination of states, each with a certain probability.

Remember the double-slit experiment of Thomas Young? In that experiment, a photon appears to pass through two slits at the same time - we can explain this by describing the particle as a wave, but we could also illustrate the phenomenon with other words like the particle was in a superposition of **all** the possible paths. Add up all those paths, and you get something that looks like a wave.

In a sense, superpositions are just a way of talking about quantum uncertainty, which means that **nothing is well-defined in this microscopic universe**. In quantum mechanics, it was observed that we cannot know any property of a particle with 100% certainty not because of a lack of sophistication of our measurements, but because quantum mechanics simply doesn't allow it.

So if we can't say if a particle is over here or over there, then for all practical purposes it is in both places at the same time... and many others. It seems a little bit fuzzy right? It's completely normal.

When an electron is moving through space behaving like a wave with no specific location, we can state that it's in a superposition. But when we actually observe it splatting against the wall, we only see that splat in only one spot - making the observation appear to have destroyed that *superposition*. **In short, the particle transformed itself from a wave-form entity to a specific tiny object located in space.** This is because we detected it, and observed it.

The electron had to choose one spot to interact with the wall in. Every time we observe a particle, we're intrinsically ruining a perfectly good superposition. This has led us after numerous experiments the following question: What does it mean to observe an experiment? Do we have the power to change reality with our eyes? Do we create it? Is it our consciousness in a way involved in the manifestation of the phenomenon observed? The prevailing interpretation at the time was that any isolated quantum system is always in a superposition state up until the moment it interacts with the world outside.

You can conduct an experiment inside a sealed box and as long as the experiment involves quantum uncertainty, the result is in a superposition mode until the box is opened and observed. In discussing the situation with his pal Einstein, Schrödinger came up with the example of the experiment of the cat to illustrate the absurdity of this idea.

Now, there are many variations of this thought experiment but the main concept remains intact. What if, said Schrödinger, we put a healthy and sleepy cat in a box with a device which in its interior is placed a strong radioactive material that has some probability of decaying between now and the end of the experiment?

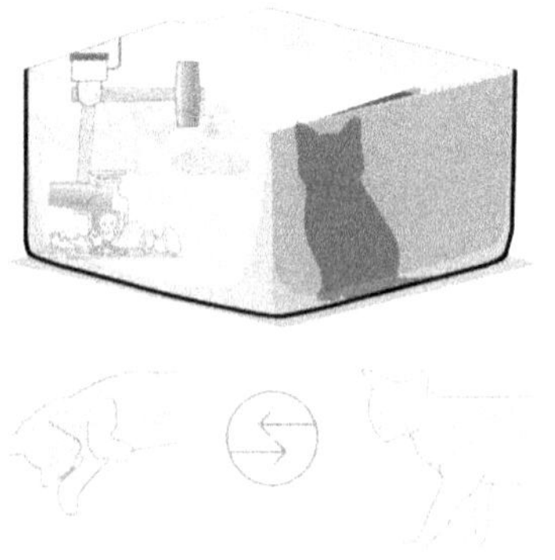

In other versions is told using poison as an example. But anyway, before opening the box we don't know the state of the radioactive material - as far as we're concerned is in a superposition of intact and decayed which means the cat in the box is both dead and alive.

According to this hypothetical scenario, opening the box is the only thing that settles the cat into one state or the other.

The cat is both happily alive and dead, it's a heads or tails scenario, a spinning coin that doesn't stop until you decide to watch.

Both Einstein and Schrödinger considered this to be utterly nonsense - but exactly how quantum states transition from superpositions to something more definite is still an open question as is why particles can be in superpositions, but large objects like people, can't.

Paraphrasing what Neil Tyson once said:

The popularity of the Schrödinger's cat is due and related a little bit to what people have called the observer effect, which consists in the changing of the aspect of reality because it is observed by someone with consciousness.

It's unfortunate that someone has called it the observer effect, because all the New Age movement, that we have seen take hold in recent decades, and other people who are (most of them) scientifically illiterate, were like:

"It's your consciousness that affects what you're observing, it's your mind that shapes the world", through the power of an undefined consciousness field.

Let's cut through all of this in a simple way: the reason why we can see each other, our faces, and the world in general, is because of light.

The vision is made possible precisely because light radiates everything, all around us, the objects, nature, the faces of our beloved ones, etc, etc. All of this is because of these radiations. Some of these radiations are bounced off the objects, some of them go through our natural optic system and Boom! We can see each other.

Light carries energy, every photon of light that strikes our body carries energy, most of them are reflected, some of them deflected and others get absorbed.

So, here's the thing - if you get tinier, tinier and tinier and you're not longer a macroscopic being but a microscopic entity - there's a particle size below which in shining light on you, that light will hit you so hard, that you pop into another location. In order to see you in this new configuration of yourself we need light, but in the precise moment we use light (of whatever part of the spectrum) and hit you, you've been already swept away.

So you are there, you are not there? If yes where precisely? We will never know because in order to see/ locate you, we need to hit you with something that makes you not there in that precise spot once you are exposed to! Which necessarily must be a form of radiation/light in the electromagnetic spectrum. And since it happens right at the moment you're exposed, (not the moment we see you), we will never know what you were doing! This of course if you are small enough for that light energy to affect you in that way.

So it's not so much an observer effect but it's much more a ***measurement effect****, it has nothing to do with the consciousness, it's just the device to measure you and the system of quantum mechanics that doesn't allow it.*

In short the fable of the Schrödinger cat together with the impossibility of investigating the changes in reality to the full, reveals nevertheless the beautiful aspect of quantum mechanics, which is without any presumption, a huge cultural revolution in the history of humanity.

It has forced us to rethink what is reality from scratch at its most fundamental level.

4.2 The Equation of Mr. Erwin

Now, let me introduce something that may be perceived as scary to the ordinary mortal, a couple of hieroglyphs that are essential to illustrate in this chapter. The Schrödinger wave equation.

The Schrodinger wave equation is a mathematical expression that aims to describe the energy and position of the particles in space and time, taking into account the matter wave nature inside an atom.

It is based on three considerations, which are:

- Classical plane wave equation;
- Conservation of energy;
- Broglie's hypothesis of matter-wave.

But before we continue, let's take a quick and brief incursion into what Broglie had to say about the atom. It will be useful to us.

Louis De Broglie was a renowned French physicis who hypothesized that particles, such as electrons or other subatomic particles, exhibit wave-like properties as well as particle-like properties. He applied this concept for every existing particle not only for photons like Bohr did. This hypothesis was based on the famous wave-particle duality principle proposed by Albert Einstein and Max Planck.

Louis De Broglie

Broglie states that for every momentum-carrying particle, there exists a wavelength that is inversely proportional to that momentum. The formula that gives this wavelength, also called the De Broglie wavelength, is:

$$\lambda = h / p$$

where λ is the wavelength, h is the Planck's constant, and p is the momentum of the particle. This hypothesis was groundbreaking, as it provided a new perspective on the behavior of matter at the microscopic level. It suggested that particles not only possess particle-like properties, such as mass and position, but also exhibit wave-like characteristics such as interference and diffraction. So the whole atom has a dual nature. He explained that the reason we don't see matter as a wave is that it has such a large mass compared to Planck's constant that the wavelength is so small that we don't see it.

One of the key implications of De Broglie's hypothesis was the explanation of the electron diffraction phenomenon. It explained how electrons, when passing through a narrow slit or encountering a crystal lattice, could exhibit diffraction patterns similar to the interference patterns observed with light waves. This discovery laid the foundation for the development of electron microscopy and the understanding of atomic and molecular structures.

The original goal of developing Schrödinger's equation was to better understand how these particles behave. It describes in great detail the shape of the probability waves or wave functions that govern particle motion.

Additionally, the equation explains the ways in which these wave-particles are affected by things outside of themselves and incorporates the idea of energy conservation, which provides information on the behavior of an electron bound to the nucleus.

This equation forms the cornerstone of quantum mechanics and it comes in various forms. However, in its simplest form, it looks something like this:

$$i\hbar \frac{d}{dt} |\Psi(t)\rangle = \hat{H} |\Psi(t)\rangle$$

Oh, my poor eyes...

Don't freak out, by the end of this section you'll have at least a basic understanding of what each of these symbols represents and what they're trying to tell us.

Let's start with this symbol here $|\Psi\rangle$

The bit in the middle is the Greek letter *"psi"*. The line and the pointed bracket are put around the *"psi"* to let us know that we're looking at a quantum state. The quantum state is just a mathematical description of a quantum system.

What is a quantum system? In simple terms, it's any given situation and/or information at our disposal for study. For example, if you're studying how a helium atom behaves, then this atom is your quantum system, as well as any other extra information that we need such as "where precisely is this helium atom?", "what's happening in its surroundings?", "how do the particles and electrons interact with each other?", "what happens if we place it in a magnetic field?", all of this is our observed quantum system. So in essence, a quantum state just tells us the state of the quantum system. Very simple.

Now that we know this, we can deduce that the rest of the elements in the Schrödinger equation are linked and are going to affect the quantum state inevitably.

Now let's look at this bit now : $i\hbar \frac{d}{dt}$

pronounced "***i***" "***h bar***" times "***d** over **dt***"

This part is very simple to understand. Firstly it's important to note that both *'i'* and *'hbar'* are just constants. They're just numbers, like the value of PI in the circumference of a circle, PI is just a number.

The digit *'i'*, is the square root of negative 1. Now if you haven't studied complex numbers at school, then you'd think this is something that is not quite possible - we can't take the square root of any negative number! Well, that's not technically true, we can! But let's not delve too deep into that now.

Instead, let's look at this bit: *'d/dt'*, they just measure how fast something changes with respect to time. Do you remember the formula of speed? Same here.

$$\text{Speed} = \frac{\text{Distance}}{\text{Time}}$$

We know that the speed of something is how far it travels in a given unit of time or how much distance it covers every second/minute or hour.

$$i\hbar \frac{d}{dt} |\Psi(t)\rangle$$

So '***iħ***' times '***d/dt***' of the quantum state, translate in: we got some number (*i*) to which a constant is applied (*ħ*) multiplying the speed of changing (*d/dt*) of that given quantum state (Psy)$|\Psi\rangle$.

We covered most of the equation now, but we need to come to the juicy bit. The "*H*" with a hat on top.

$\widehat{H}$ This is known as the Hamiltonian operator. But why this hat, what represent? In quantum mechanics, there are these things called operators distinguished by this little hat on top of their symbol and they basically "operate" on quantum states, they do things to the quantum states.

The Hamiltonian operator is a pivotal component because is linked to the total energy of the system, it often takes into account kinetic energy, potential energy, thermal energy, etc. all of these different types of energy.

So essentially what Schrödinger's equation is trying to say is that the quantum state (Ψ), is changing with time in a very specific way, depending on the total energy of that determined system.

This equation therefore allows us to **predict** how a quantum system should behave if all the energy terms and factors in the system are mathematically written down.

Nevertheless, there's a problem. For systems larger than a few particles, becomes basically impossible to predict anything really, really quickly. We need numerical methods, and we need computers doing it for us - but at this point in time, even our fastest computer can't handle the analyses of systems that are larger than a few dozen atoms! And so this is the limitation we face in our day and age with the computing power that we have.

To give you a better sense of what we're talking about, let's take for example a helium atom, a very simple one - it has 2 protons, 2 neutrons, and only 2 electrons. For simplification let's make some assumptions: let's assume that the nucleus doesn't move, is stationary, and electrons are the only ones that move around. In reality, of course, that's not true, the nucleus does move, and the particles in the nucleus do jiggle about but let's not go into that just yet.

Even in this huge simplified situation, the Hamiltonian operator, which consists as said of all the energy terms in the system would appear to us as a wall full of calculations and variables very difficult to manage. It would contain terms such as the kinetic energy of the first electron, the kinetic energy of the second electron, the electrostatic attraction of the nucleus to the first electron, the same with the second electron, the electrostatic attraction between the two electrons, and so forth.

And remember this oversimplified model doesn't even take into account any interaction of every part in its nucleus - as well as, the proton spin, the neutron and electron spin, etc.

So, even though we have this super powerful equation that could give us deep insights into how the micro-universe works, we have to come up with increasingly clever and complicated ways of dealing with this equation without expecting our computer to crash.

But not all hope is lost, in many cases there are workarounds to this problem, have you ever heard about quantum computers? Well, we'll talk about them later.

4.3 The Schrödinger's atomic model

In this journey through the history of the atom, we now find ourselves at a point in which the reference model can now be defined, from Bohr onwards, as quantized. The model proposed by Schrödinger is 100% a quantum mechanical model, and that is perhaps the most abstract yet the most profound model of the atom that we have.

Unlike the previous models, such as Bohr's model, which depicted electrons as discrete particles orbiting the nucleus in fixed paths, Schrödinger's model introduced the concept of electron **wave functions**.

According to this model, electrons do not have precise positions or trajectories but rather exist as probability distributions within certain regions of space called orbitals.

Orbitals? You meant orbits! Ok, let's start with the basics, so as not to confuse these two essential terms.

In classical physics, an **orbit** refers to the path followed by a celestial body, such as a planet, around another object under the influence of gravitational forces.

These orbits are well-defined, predictable, and can be represented by simple and stable elliptical shapes. However, in the quantum world, the concept of orbits undergoes a significant transposition. It's the movement of electrons around the atomic nucleus in a manner similar to planets circling a star. This classical depiction though, fails to consider the inherent wave-particle duality of electrons and is insufficient to explain their behavior at the atomic level as we saw in the section dedicated to Bohr.

Orbitals, on the other hand, are the regions in space where electrons are most likely to be found. They are described by mathematical functions known as wavefunctions, which provide information about the electron's energy, position, and spin.

Unlike classical orbits, orbitals do not have well-defined boundaries or fixed paths.

In short, the main differences between orbitals and classical orbits can be summarized as follows:

- **Nature**: Orbits are classical paths with well-defined trajectories, whereas orbitals are probabilistic descriptions of electron locations.

- **Predictability**: Orbits can be precisely determined and predicted, while orbitals provide a statistical framework for determining electron distribution.

- **Shape**: Orbits follow predictable elliptical paths, while orbitals have diverse shapes and are defined by mathematical functions.

- **Quantum Numbers**: Orbits do not have quantum numbers associated with them, whereas orbitals are characterized by a set of quantum numbers that determine their properties and energy.

So in this new version of the atom, we have a probabilistic model that describes the likelihood of finding an electron in a certain position and it does not describe the exact path or motion of the electrons.

This is an example of an orbital, you can see that the density of the dots is much greater around the nucleus in the middle and it decreases as you go away from the nucleus.

This orbital diagram implies that it is more likely to find an electron closer to the nucleus and that the probability of finding electrons decreases as the density decreases away from the nucleus.

The Wave Function

The wave function, denoted by the Greek letter psi (Ψ), as we've seen earlier, is a complex-valued function that varies with position and time. It contains all the possible information about the particle's position, momentum, energy, and other physical properties.

The relevance of the wave function to this probabilistic model is that when you square this wave function it provides the probability density of the electron, which is what we saw in the orbital diagram.

This graph provides here is a visual representation of the square of the wave function - in other words the electron probability of three different orbitals, the 1s, the 2s, and the 3s orbital.

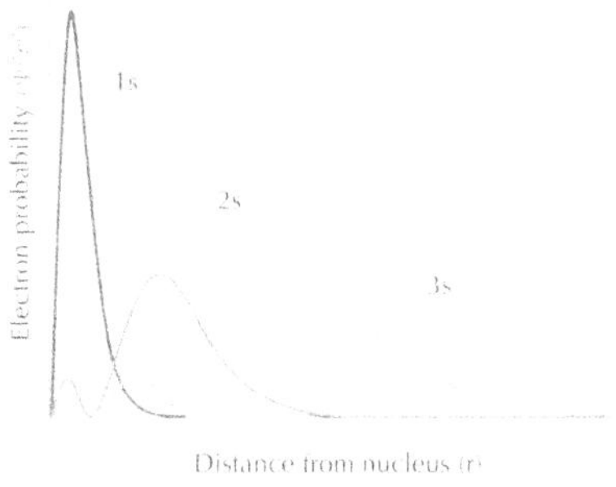

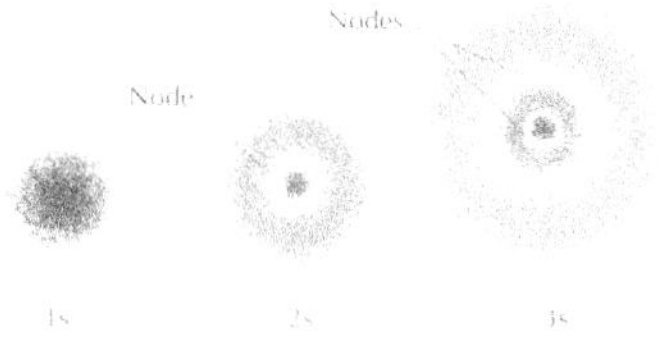

You can see that the probability is not constant and uniform, it changes throughout each orbital and the probability is also different between different orbitals.

We can see a direct comparison of what the probability density diagram looks like in relation to the graph shown. The 1s orbital is surrounded by the 2s orbital and the 2s orbital is further surrounded by the 3s orbital. You can clearly see that between orbitals there's a white concentric region where is impossible to find electrons, these are known as **Nodes**.

These regions of zero probability account for the quantization of the energy states that were discovered by Planck and then subsequently explained by Bohr. Let's remember that quantization of the energy state means that the energy of an electron's orbital must be multiple of a specific value and the nodes are forbidden areas where you cannot find electrons as the energy in these regions is not a multiple of that small value.

Quantum Numbers

The quantum parameters or as they are called in this context "*numbers*" are associated with Schrödinger's atomic model as they describe the quantum state of the electron. There are a total of four quantum numbers, each quantum number describes a different **aspect** of the quantum mechanical nature of an electron. According to Pauli's exclusion principle, no two electrons can share the same quantum number. That means that in a given atom, every electron has its unique set of four quantum numbers.

<u>The first quantum number is the:</u>

Principal quantum number (n) which relates **size and the energy** of the orbital - whether it's 1s orbital, 2s orbital, 3s orbital, etc. The energy is also related to the distance of the orbital away from the nucleus;

<u>The second quantum number is the:</u>

Angular quantum number (l): this relates to the **shape** of the orbitals, which are not all a perfect sphere as you might imagine;

<u>The third quantum number is the:</u>

Magnetic quantum number (ml): this relates to the **orientation** of the orbitals, aligned to three imaginary axes (x,y, and z);

The fourth quantum number is the:

Spin (ms) - it determines the magnetic spin property of electrons, which can either be Up or Down. In each orbital, there can be a total of two electrons, with one electron taking the up spin and the other taking the down spin.

The size energy of the orbital increases with its distance from the nucleus, so the 2s orbital is further from the nucleus and therefore has higher energy compared to the 1s orbital, and the 3s orbital is further away than the 2s orbital, therefore it has greater energy compared to the 2s and so forth.

On the other hand, the angular and magnetic quantum number determines the shape and orientation of the orbital. But so far we've only looked at the "s" orbital. However, orbitals can take on different shapes such as "***p***" orbitals, "***d***" orbitals, and "***f***" orbitals. Amongst each type of shape of orbital, you can have different orientations.

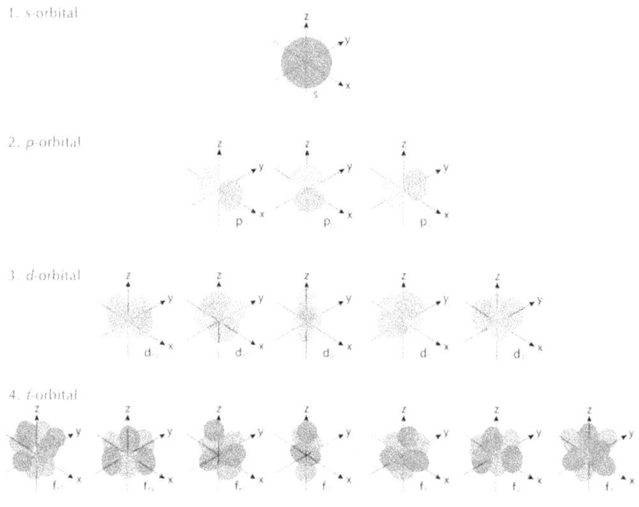

Examples of the different kinds, shape and orientation that an orbital can have

For example, the "***p***" orbitals can take on three different orientations where the plane is either on the x, z, or y plane - "***d***" orbitals can take up 5 orientations, and "***f***" orbitals can take on 7 different orientations.

4.4 Understanding Electron Spin

Let's begin to address this topic by illustrating a couple of points that could be defined as crazy and inconsistent characteristics when comes to describing the electron:

Electrons are particles but are also waves;

Electrons do not orbit but buzz around in specific areas;

We don't know where they specifically buzz in that area;

We don't know what they're going to do.

And in addition to this, they DO NOT SPIN!

What? That's not true!

Everyone knows that electrons spin! Well, to be exact it appears that there exists some kind of angular momentum (aka rotation) but it's not a spin in the classical meaning of the term. So what kind of witchcraft is this electron?

Let's assume for a second that an electron is a solid sphere, like a marble for example, which is not a far-fetched idea, right?

If the diameter of the electrons is 10^{-19} meters, then for them to spin with an angular momentum of one h-bar (ħ), it was calculated that their outer edge would need to be traveling at nearly a million times the speed of light which is inconceivable.

Electrons are considered fundamental particles, meaning they have no substructure and they do not have a well-defined size or shape. However, scientists have developed various methods to estimate and discuss the "size" of an electron.

It is important to note that the term "diameter" used in this context is somewhat metaphorical, as the electron is not a sphere. Instead, it represents its spatial extent, more than its actual shape, or in other words the region around is most likely to find its "expression". To not get tied up in abstruse terms nowadays the electron is described as a point-like particle, which size is so infinitesimal and mass so tiny that is considered negligible.

So, due to this out-of-charts speed that the electron should have, and its negligible dimensions, the only possible conclusion is that it's not spinning. Point-like things by logic cannot spin. To do that the object in question should have a mass and an axis.

To be clearer, particles have all sorts of properties like mass, charge, position, energy, etc. and spin is one of these properties - we can sort these properties into two categories:

- properties the particle have because of what they are: like mass and charge;
- properties the particles have because of what they're doing: like position and energy.

Spin in quantum mechanics, against any common reasoning, falls into the first category. It's a type of angular momentum a particle has just because is that particle and for no other reason. It's in its intrinsic nature - which contradicts the reality of our macro-world where the spinning of an object is due to what it does.

So at this point, it shouldn't even be called spin! Intrinsic angular momentum is a more accurate description, but it's not nearly as catchy. The reality is that this universe is just so unfamiliar to us that there isn't a better word for it.

So does this spin exist or not in electrons? How do we know it?

Short answer: Yes, exists this intrinsic angular momentum and we know it because of magnetism!

Any electrical engineer will confirm that any charge with any kind of momentum is going to manifest a magnetic field, which means it's going to be affected by another magnetic field. It was observed that some minerals such as magnetites (where magnets come from) manifest an intrinsic magnetism due to the alignment of the momentum of their particles.

In the same fashion other alloys of metals, can be magnetized if put in contact with strong magnetic fields, and this happens because the magnetic field tends to align the momentum of the particles of the metal that undergoes such influence.

That's the reason why we know there's a sort of sub-atomic movement that allows the manifestation of a magnetic behavior at the particle level and it was observed that this magnetic behavior is affected by other external magnetic fields.

Okay, but if it's not a classic spin that creates the magnetic field around the particles, then what does this momentum actually consist of?

Get ready my dear reader, because what comes is something a little bit tricky to grasp, but it will blow your mind. We are dealing with topics whose subjects undergo dynamics that don't follow the rules of our daily macro-world - therefore it can be hard to create a mental image.

4.5 The spinors

As we saw at the beginning of this chapter, Schrödinger's revolution led to the acceptance that the atom is a quantum model whose electrons are in a superposition state, which means that the intrinsic characteristics of the electron are in superposition too. Yes! including their spin! Because remember spin is a property of what electrons are, not a property of what electrons do.

The spin in electrons and other particles (including neutrons and protons) are quantum mechanical "objects" that manifest themselves in a duality state (0 and 1). It is conventionally called Up and Down, but as we said due to this unusual configuration the literal meaning of up and down has little sense (there's no up and down per se, it depends on your reference point), we can call them Left and Right or Jimmy and Bob, it doesn't matter. What represents here Up and Down or 0 and 1, is the intrinsic polarity of these microscopic "diapoles" we call particles.

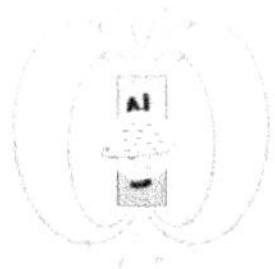

They manifest a magnetic field just because they exist in their inevitable quantum state, even neutrons with no charge manifest a magnetic field.

It was Wolfgang Pauli, an Austrian physicist in 1928, who introduced the concept of **Spinors** to describe the behavior of quantum particles when comes to spin.

Wolfgang Pauli

One of the most intriguing aspects of spinors is their transformation under rotations. They are exceptionally weird and cool and they really deserve their own book, but let me illustrate a couple of things to give you a taste.

They describe particles that manifest very strange "rotation" properties.

The concept of spinor introduced by Pauli can be visualized as a mathematical object that represents the quantum state, meaning that in order to return to its initial position the particle needs to complete twice a turn.

So we're admitting that it's involved a sort of rotation but with different features. To return to its initial position the particle needs to "rotate" 720°, it's not sufficient 360°.

An example of spinor-like behavior could be pictured as an attempt to rotate a mug in your hand without letting it go. In doing that, your arm gets twisted, but a second rotation of your wrist untwists your arm.

Another way to see it is as a sphere or cube fastened to some imaginary walls with ribbons; if we spin it around a full circle, the sphere is back where it was, but the ribbons are twisted from where they started. The entire system returns to its initial condition when, astonishingly, we spin it around again (not in the opposite direction, but in the same one).

One revolution generates a twist, and two rotations return it to normal. So, imagine the electron as being linked to every place in the universe by invisible strands. Accordingly, the spinor wavefunction's phase is angle-dependent, and a full rotation causes it to be out of phase with respect to its initial value.

So even if the electron is point-like, its angular momentum can arise from an extended, though still tiny region.

Crazy right? I strongly recommend watching a couple of videos on YouTube to watch how mesmerizing their motion is, like this one:

This crazy idea came up after multiple mathematics calculations, these calculations were integrated into Schrödinger's equation which was lacking details about the spin of particles. In three-dimensional space, a rotation of 360 degrees would bring an object back to its original orientation, but in the quantum realm, a similar rotation may result in a different quantum state. This phenomenon is known as quantum indistinguishability, and spinors capture this intriguing feature.

"Spin Up"

So when we say a particle has a spin up, it means that its spin is aligned in the direction of an external magnetic field, conventionally represented as the positive z-axis in an imaginary graph. This alignment occurs when the particle's spin angular momentum points in the same direction as the magnetic field's angular momentum. It is often depicted as an arrow pointing up.

In quantum mechanics, the spin of a particle can take on specific discrete values. Typically, the spin is measured in half-integer units, such as 1/2, 3/2, and so on.

The electron as illustrated possesses a spin of 1/2, which means it can exist in two distinct spin states.

"Spin Down"

Conversely, when a particle has a spin down, its spin is aligned opposite to the direction of the external magnetic field, represented as the negative z-axis. Similar to spin up, spin down is associated with an arrow pointing down. It signifies that the particle's spin angular momentum points in the opposite direction to that of the magnetic field's angular momentum.

It's important to remark once again that the terms "up" and "down" are merely conventional labels used to describe these distinct spin states. They do not imply any physical orientation in space but rather describe the orientation of the spin relative to external stimuli like a magnetic field.

The significance of spin up and spin down lies in their role in determining the behavior of particles in various quantum phenomena. For example, when particles with opposite spins interact, they can exhibit unique characteristics, such as entanglement, which is a fundamental aspect that we will soon talk about.

Thus, summarizing electrons don't whirl, they're engaged in a far more fascinating dance. If we think about it, the motion we call spin may hold the key to understanding the structure of matter and, by extension, reality. And it's true, it appears that the subatomic fabric of spacetime contains these bizarre and peculiar tiny knots. Fascinating...

A bravo to Mr. Pauli! I think as science enthusiasts applause is due to him.

4.6 Life of Schrödinger

Let's now return to our chapter's main character for a second as he is quite an interesting figure.

Erwin Rudolf Josef Alexander Schrödinger, one of the most famous names in quantum physics, was born in Vienna on August 12, 1887. He was raised in a religious household, his father was

Catholic and his mother was Lutheran, despite this environment, he always considered himself an atheist.

Though atheist, he had a strong interest in Eastern religions and pantheism and he used religious symbolism in his works, he also believed his scientific works were an approach to the godhead albeit in an intellectual sense.

Schrödinger attended the University of Vienna to study mathematics and physics when he was younger. Retired physicist Franz Exner mentored him while he earned his Ph.D. Since he was a young boy, Schrödinger had a profound interest in philosophy and color theory; in fact, he once stated in his lecture "*Mind and Matter*" that "*The world extended in space and time is but our representation*" - a statement that reflected his passion for Schopenhauer's writings.

In the 1910s, Schrödinger became acquainted with the works of Planck, Einstein, and others, but wasn't terribly interested in giving up classical physics just yet. Meanwhile, he decided to leave Germany since he disliked the Nazis and the anti-semitism.

All changed when he developed tuberculosis and spent some time in and out of sanitoriums regaining his health. He became famous for his "feud" with Einstein, due to Schrödinger's so-called failure to embrace the fundamentals of quantum physics as early as some of his colleagues.

While convalescing, the scientist started to warm up to the ideas placed before him, especially quantum theory. He was fascinated with the possibility that the electromagnetic spectrum could have differing effects on a wide variety of elements and wanted to discover if he could find a universal way to predict the behavior of electrons.

Schrödinger played around with these thoughts and came to the conclusion that the only way to do this would be to be able to predict the nature of radiation itself.

While he couldn't find a way to do this, the articles he published proposing these concepts opened the door for a new brand of theoretical physics that led to a few concrete advancements, such as wave mechanics and an interesting new atomic model.

According to James D. Watson's memoir "*DNA, The Secret of Life,*" Schrödinger's 1944 book "*What Is Life*" sparked Watson's interest in studying genes, which ultimately led to the discovery of DNA in 1953. The book discusses entropy and the idea of complex molecules with genetic codes for living organisms.

His extremely unconventional marriage was made public when he officially emigrated to Ireland with his mistress, wife, and any children he had with either of them. While he was a resident of Dublin until his retirement in 1955, he maintained a lifelong interest in the Hindu Vidhata school of thought, which informed his ideas about the nature of reality and the possibility that our own consciousness is really a reflection of universal unity.

In 1956 he returned to Vienna and at an important lecture during the world energy conference, he refused to speak of nuclear energy because of his skepticism about it, and instead, he spoke on philosophy at length until his audience protested.

Erwin Schrödinger died of tuberculosis in 1961 and was laid to rest in his native Austria. His wave equation is engraved above his tomb, and his wife was buried with him after her death in 1965. No word on what happened to the mistress.

Schrödinger was fabulously gifted, delightfully weird, and a tremendously impactful person in the world of quantum physics.

HEISENBERG AND HIS UNCERTAINTY

In this chapter, we will explore an intriguing concept known as Heisenberg's Uncertainty Principle. This principle, formulated by the German physicist Werner Heisenberg in 1927, implemented our understanding of quantum mechanics.

He is widely recognized for his groundbreaking work on the development of matrix mechanics and the formulation of the uncertainty principle.

W. Heisenberg

Heisenberg's fascination with physics began at an early age, and he pursued his passion by studying physics at the University of Munich and later at the University of Göttingen. Under the guidance of influential physicists such as Max Born and Arnold Sommerfeld, Heisenberg's talent and dedication quickly became evident.

The Uncertainty Principle, his crowning achievement, was born out of his quest to comprehend and characterize the actions of subatomic particles.

It is physically impossible to measure two physical variables, like location and momentum, with absolute precision, as he proved. A probabilistic understanding of quantum physics was bolstered by this principle, which essentially contested the classical concept of determinism.

Despite facing criticism and controversy, Heisenberg persevered and continued to contribute significantly to the field of quantum physics throughout his career. He received numerous accolades for his pioneering work, including the Nobel Prize in Physics in 1932.

5.1 The Mathematical Formulation of Heisenberg's Uncertainty Principle

As we were saying, this principle of uncertainty is a mathematical expression that is expressed in wave functions which carry in itself a really baffling concept!

It basically tells us that there is a fundamental and universal limit to how much we can know about pairs of quantities known as "conjugated variables". Conjugated, simply means, related to each other or linked together in a specific way. Now, most commonly the uncertainty principle is described using the quantities of Position (mathematically expressed with an ***x***) and Momentum (mathematically expressed with a ***p***) which is roughly the speed of a particle as viewed previously.

With these quantities told, the uncertainty principle says that the more we know about one the less we can know about the other - more specifically, the principle tells us that if we take our uncertainty in a position that we'll call Δx and multiply by the uncertainty in momentum Δp, then this product multiplied together is greater than or equal to $\hbar$ (h bar) the Planck's constant divide by 2.

$$\Delta x * \Delta p \geq h/2$$

So translating in a more human language, what this formula is trying to explicate is that when we multiply two uncertainties together this has to be greater than or equal to some number.

In other words, the product has to be at least this number (the Plack's constant).

What it means is that we cannot take the two quantities Δx and Δp and make them both as small as we'd want them to be (we want both uncertainties to be very small), the more the uncertainty is low the more accurate we can be with our measurements.

If we want to make one of them small, then the other quantity has to get larger, so when we multiply them together it has to be at least $\hbar/2$.

Why is that so significant? Because these uncertainties represent uncertainties in our measurements of the position and momentum of a particle. In the macroscopic world, the position of an object in a determined space is a problem that simply doesn't arise, but at the subatomic level, it definitely does.

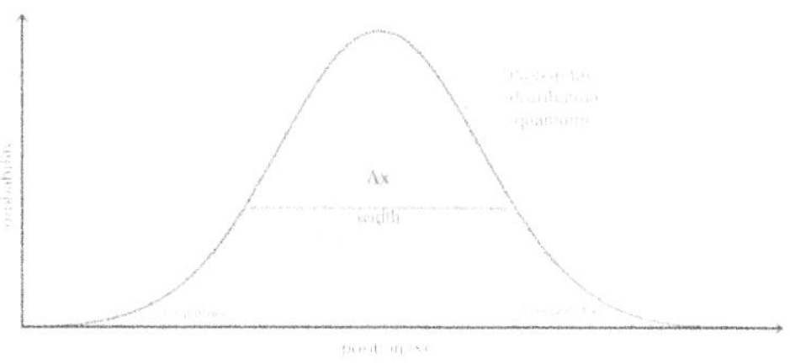

Before measuring the position of the particles we must acknowledge that we have a probability distribution, and surrender to the fact that knowing the exact point precisely is almost impossible. Usually, the area that coincides with the peak of the wave function is the region where the particle is most likely to be found - in other terms, there's a very high probability that the particle is where we measure it to be classically. However, it also has the probability of being somewhere else and it is the width of this probability distribution that defines the uncertainty in our measurement of that particle's position. In other words, the uncertainty is how uncertain we are about our measurement.

So, we cannot make both quantities Δx and Δp, arbitrarily small, we can't even make one of them zero that is, to know exactly where the particle is at 100%. Having that certainty would result in the momentum of the particle being infinite which translated means knowing absolutely nothing about its momentum or speed. It's something that just never happens.

5.2 How measurements occur

Let's consider we have a particle. It's just doing its things as a particle of buzzing around in space, sometimes annoying its peers, and living its duality. How would we measure the position and speed of that particle? We can't just stick a ruler beside it, it's incredibly small, it's a particle!

What we have to do is to fire a light to the particle. Light (to be more precise any radiation) is the only medium that allows us to enter into the mechanism of reality, is the natural phenomenon that allows us to actually see.

Light can bounce off the particle and give us information about the particle's position and momentum - now the issue here is that at this scale, quantum effects start to occur. The light we fired at it, has quantum properties, it has wave-particle duality due to its component which is the photon.

The interaction between the fired light and the particle introduces a disturbance in the momentum's particle, known as the "*quantum back-action*."

Quantum back-action arises from the inherent wave-particle duality of matter and the probabilistic nature of quantum mechanics. When photons interact with a particle, they transfer both energy and momentum to it. Consequently, a **destabilization** of our observation frame occurs, and the direction and speed undergo changes as a result of this interaction.

The impossibility of measuring precisely both properties arises due to the quantum back-action. The transfer of energy and momentum from the photons to the particle leads to an unavoidable disturbance in the particle's momentum.

How can we narrow the position wave function of a particle?

Let's reiterate - to understand how we can narrow or enlarge the position wave function of a particle, we must first recognize that the position wave function is essentially a mathematical representation of the probability of finding a particle at a particular position. The narrower the position wave function, the more localized the particle's position becomes, indicating a higher probability of finding the particle in a specific region of space.

One approach to narrow the position wave function is through the process of localization. Localization involves confining the particle to a smaller region of space, thereby reducing the uncertainty in its position. This can be achieved through various experimental techniques, such as using highly focused laser beams or employing sophisticated trapping mechanisms. By restricting the particle's movement and confining it to a smaller region, we can increase the precision with which we can determine its position.

On the other hand, enlarging the position wave function involves allowing the particle to explore a larger region of space, thereby increasing the uncertainty in its position. This can be done by providing the particle with more freedom to move and reducing any external constraints.

How can we narrow the momentum wave function of a particle?

The momentum wave function provides us with information about the likelihood of finding the particle with a certain momentum value. However, the momentum wave function is inherently broad, encompassing a range of momentum values.

In order to narrow down the momentum wave function of a particle, certain techniques and measurements can be employed. One such method is through the use of position measurements.

By precisely determining the position of a particle, it becomes possible to gain information about its momentum as well. To narrow the momentum wave function, one must sacrifice some knowledge about the particle's position. By making a less precise position measurement, it becomes possible to obtain a more accurate measurement of momentum.

Another way to narrow the momentum wave function is by using specialized instruments and techniques. For example, a particle's momentum can be determined through the process of scattering, where the particle interacts with other particles or objects. The resulting change in trajectory or velocity provides information about the momentum of the particle.

Additionally, advanced experimental setups such as particle accelerators can be used to manipulate the momentum of particles. By controlling the energy and trajectory of particles, it becomes possible to narrow down their momentum wave function.

It is important to note that while the momentum wave function can be narrowed, it can never be completely eliminated. This is due to the inherent uncertainty in quantum mechanics, as described earlier. The principle sets a fundamental limit on the precision with which certain pairs of physical properties, can be simultaneously known.

And this is how we can understand Heisemberg's uncertainty principle, it's a fundamental property of the universe. This uncertainty has nothing to do with our current technology or our intelligence level - it's just a property of the universe and according to quantum mechanics, there's no way to get around that.

But let's be clear the uncertainty principle isn't saying that we're always uncertain of everything. It just saying that we can't be particularly certain about more than one thing at a time.

Sometimes could be useful to picture the wave like it's made of a swarm of bees - when you let them spread out they will probably just hang around and be chill, on the other hand, if you try to contain them all in a really small space, they will most certainly be buzzing around furiously. Based on the environment they are in, you will have two different scenarios, which in the quantum realm translates to two different data sets with two different precisions.

FURTHER QUANTUM PHENOMENA

As you may have understood, the quantum world is very vast and there are a myriad of phenomena which for each of them would deserve an extensive book.

With chapter 5 we have concluded the historical overview of the evolution of the atomic model and the most important and central characteristics of particles at the quantum level.

Here in this chapter, we'll see together other quantum phenomena that arouse more interest in the public and whose implications have revolutionized the field of study of matter itself.

Let's dive in.

6.1 Quantum Entanglement

"Spooky action at distance" is how Einstein himself described this phenomenon, quantum entanglement is a marvel in quantum mechanics that involves the correlation of quantum states between two or more particles, even when they are separated by large distances. It is a concept that defies classical intuition and has perplexed scientists since its discovery.

The EPR conundrum, a 1935 article by Einstein, Boris Podolsky, and Nathan Rosen, was the first to suggest the idea of entanglement. A more thorough description and mathematical formalism for entanglement were, however, given by Schrödinger.

What is it all about?

When two or more particles grow so linked that their states cannot be stated separately, we say that they are entangled. Quantum superposition and energy exchange are two of the many pathways that might lead to this entanglement.

When two particles, A and B, get entangled, their separate quantum states combine to form a single state that includes all of the characteristics of both particles.

This means that the properties of one particle are instantaneously connected to the properties of the other particles, regardless of the distance between them.

The classic example would involve two electrons - electrons, as you already know, have a quantum property called spin. The spin can have two values, it's like a sort of quantum heads or tails - spin up and spin down.

If the electron becomes entangled, then this may create a situation where those two spins are correlated such that if one of the electrons has spin up, then the other one must have a spin down.

Picture this: let's suppose we have two gloves, a left-handed and right-handed glove, let's send them out to two people on different sides of the world, and these two people, are going to meet each other shortly, they're called Ariel and Bob.

They know that these gloves are a pair and as soon as Ariel receives her glove, opens the package, and finds she has the left-handed glove, then she knows right there that Bob must have the right-handed glove because there's this correlation between them.

However this is the complication in quantum mechanics: unlike gloves (with their fixed features), particles have their properties including their spin, not in a fixed state, they change all the time. They only have a property with a fixed value, when we observe it.

So for these two entangled electrons, if we think of sending those out to Ariel and Bob, they will arrive in the hands of both people with an unsettled orientation of their spins, all we can say in this case is that those spins are correlated.

As soon as Ariel measures her electron and finds it has spin up, then Bob's will have spin down (and vice versa). But that spin wasn't determined until Ariel measured it. This is where Einstein felt there was a problem with entanglement because it seemed to indicate that the act of Ariel measuring her electron spin somehow instantaneously affected Bob's spin. It seems to be a sort of "spooky" transmission as Einstein said, some kind of influence to Bob's spin to make sure that it was spin down.

Too much crazy to be accepted, Einstein said that there has to be some alternatives to this, because "spooky action at a distance" shouldn't be allowed in physics, it violates the principle of locality - Einstein has shown that it's impossible for any signal, any information to be transmitted faster than light. Therefore, it's impossible to have this instantaneous "action at distance", there has to be some time for a signal to span the distance.

Another suggested alternative was that all along, these two electrons had some inherent property that somehow fixed their spins already, it's just a property that we couldn't measure, some "hidden variables". These hidden variables, if they indeed existed, would provide an underlying explanation for the seemingly random and at the same time coordinated behavior of the electrons' spins.

The concept of hidden variables offers a captivating perspective, suggesting that there might be more to the nature of electrons than meets the eye. It proposes the presence of unobservable properties that govern the behavior of these particles, possibly influencing their spins in a predetermined manner. Such hidden variables, if proven to exist, would revolutionize our understanding of quantum mechanics.

So, more than communication faster than light, this is knowledge faster than light! Because the concept of communication includes the notion of space and time, (elements required to have a delivered message), in this case instead we have the instantaneous manifestation of information on the other side.

However, it is important to note that the existence of hidden variables is purely speculative at this point. Despite extensive research and exploration, scientists have yet to discover any conclusive evidence supporting the notion of hidden variables. The mystery surrounding the true nature of electron spins continues to challenge our understanding.

6.2 Quantum Tunneling

Let's continue with *quantum tunneling* another fascinating phenomenon in the field of quantum mechanics. It's a concept that defies classical physics and allows particles to penetrate energy barriers that would typically be impenetrable according to traditional Newtonian physics.

Let's break it down this way:

Imagine you're standing on one side of a mountain and you have to get to the other side, you're going to climb over the top of the mountain and come down the other side, it's the simplest way. Another way to do it is to bore a hole through the mountain and come out the other side. By doing so you made a tunnel. You make it harder, it's the shortest path but it requires a lot of effort and time.

In quantum physics, we can think of the mountain as an energy barrier to you. In addition, you need the energy to ascend the mountain, otherwise you stand where you are.

Now, let's imagine for a second that a particle needs to get to the other side of the barrier, you give the particle some energy, but it isn't enough to overcome the obstacle. How can it do it?

The particle, is not only a particle, remember? It's also a wave, and a wave is mathematically expressed in a wave function. A wave function is a probability of finding it anywhere in a volume spanned by the wave function itself. And as seen previously, you're more likely to find the particle where the wave peaks, and where the wave drops off you're less and less likely to find it.

There is though an actual likelihood (although minimal), where the particle can show up anywhere in the wave function even in the low probability places! The wave doesn't know about and doesn't care about the mountain or any barrier. It's just a matter of probability.

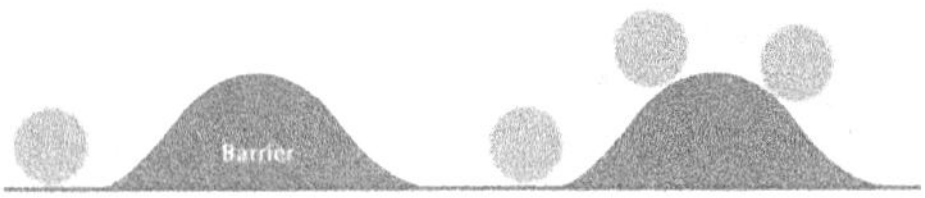

Classical Mechanics

Quantum Mechanics

So you can ask, do some of that wave could show up on the other side of the mountain? If it does, then there's a probability that the particle that is stuck on one side will just simply appear on the other side of the mountain, having never had to ascend it in the first place, because it didn't have enough energy to do it.

We can confirm that in a wave function, it exists indeed and there will always be a probability that it can show up where it was not "invited". That's called quantum mechanical tunneling.

Now, let me show you how it manifests itself but before, let me do a little digression, it will be helpful.

In the early days more than a century ago, we didn't know where you got the elements in the universe, they were thought to be already as a default setting in the Earth, but then later we learned that they're got from stars.

Theoretically, we could build them starting from smaller elements if we merge them. On paper, scientists made calculations of what would it take to merge two hydrogen atoms (the most basic atom in the periodic table). In the nucleus of a hydrogen atom we have only a proton - in this example, we place two protons that we need to merge together, if we succeed we'll get an atom of helium. That's a way to build elements from scratch.

But remember protons are intrinsically positive charges, in this scenario we need to merge two positive charges, and we know that they don't like each other because they repel. It turns out that the electromagnetic repulsion of the two positive charges can be overcome and bring the particles closer and closer to each other if we speed them up by increasing the temperature of the environment. But what temperature is required to force this and create a strong nuclear force to kick in and grab them? The answer is a billion degrees!

And there's no place in the known universe with such a temperature! Just so you know, only in the cores of stars do we have enormous temperatures but nonetheless we are nowhere near the required degrees - star nuclei have "only" millions of degrees.

Our Sun in its nucleus for example has around 15 million degrees celcius.

So the topic of the origins of the elements has just sat for decades. It was sat there as an unsolved problem until quantum physics came along.

Physicists said: we have this proton coming close to another proton and there's this barrier here that it cannot cross. But the particles are also waves, and part of that wave exists within the grasp of the strong nuclear force so there's a probability that some of these particles will merge and make helium!

The next question of the physicists was: in what percentage of collisions the wave-particle will tunnel through this barrier and end up making helium? They came up with that percentage and Baaam! Let there be light! To give you an idea it's less than 1 in a trillion encounters.

In other words, all the stars you see in the sky with their perfect burning mechanism are the result of billions and billions of fortuitous tunneling, which in turn is a phenomenon with a minimum percentage of probability. Isn't all this so freaky, considering that the stars are almost countless?

And we're actually glad that not all the encounters between protons end up with a merging reaction, the sun and all the stars would just blow up, and they would blow to smithereens in a second.

So long story short: the thermonuclear fusion in stars that generates the energy at the temperatures it does, can only happen because of quantum mechanical tunneling.

And no matter the size of the potential barrier, particles can tunnel and appear instantly on the other side because wave function just collapses and they're already there.

But let me be precise, **they don't travel the "tunnel", they were always there probabilistically, so the distance is inconsequential or immaterial if you want.**

In conclusion, due to its superposition state, the particle always has a certain degree of uncertainty associated with its position in space enabling itself to tunnel through barriers that would be seemingly impossible to cross based on their classical energy levels.

THE GENIUS OF EINSTEIN AND HIS RELATIVITY

Ladies and Gentlemen, in this book we've already mentioned him a couple of times - he is without any doubt one of the most renowned scientists in history, a German-born theoretical physicist who revolutionized our understanding of the universe. Here we present to you none other than Mr. Albert Einstein! (applause!).

He absolutely deserves a chapter in our journey in the quantum realm for obvious reasons. He sits on the Olympus of physics for his theory of relativity (and not only), which fundamentally changed the way we view space, time, and gravity. Einstein's genius and remarkable contributions to physics earned him global recognition and made him an icon of scientific achievement. His work continues to inspire and fascinate scientists and enthusiasts alike, as he unlocked some of the secrets of the cosmos and expanded our knowledge of the fundamental laws that govern reality.

7.1 A little about his life

Ulm, a city in southern Germany, was the site of Einstein's birth on March 14, 1879. Hermann, his father, was an electrochemical plant manager and a featherbed salesman. Both he and his sister Maja were cared for by their mother, who also looked to the home. Reportedly, Einstein did not begin to talk until he was two years old, and in his childhood, he had quite the personality.

He frequently rebelled against authoritative figures and had fits of intense anger. At one point, he even hurled a chair at a teacher!

The fact that Einstein wasn't the "perfect student" when he was younger is unknown to many, it was even reported that he was forgetful and daydreaming, unusual traits for someone who then became the face of the hard sciences.

He struggled in classes that emphasized memorizing and had trouble interacting with other kids. Learning settings that encouraged creativity and encouraged inquiries were ideal for him. The role of unseen forces in shaping reality particularly piqued his interest.

Actually, Einstein had a professor who warned him he "would never amount to anything."

Concerned, parents sought medical attention for Einstein when he displayed signs of a speech delay. Recent research suggests he could have been on the autistic spectrum, which would explain some of his social and communication challenges. Einstein later said that rather than words, visuals were the conduit via which his ideas and thoughts traveled. His delayed speech may have been influenced by this way of thinking as well.

He had an interesting personality and a mischievous side that often got him into trouble. One time, he decided to test the laws of gravity by launching water balloons from his bedroom window. Needless to say, his aim wasn't the best, and his neighbors became unsuspecting victims of his scientific curiosity!

Alongside his playful nature, Einstein showed great curiosity and intelligence from a young age. He would spend hours engrossed in books and experiments, always questioning the world around him.

His parents often found themselves perplexed by his never-ending stream of "why" questions. Einstein's educational path was not without its challenges. He struggled in traditional schooling due to his unconventional way of thinking.

Teachers would often scratch their heads trying to understand his unique perspective on the world.

Einstein was very gifted in the scientific and mathematical disciplines. He was reading physics texts meant for college students when he was eleven years old.
He was 12 years old when he developed a deep interest in geometry. Even at the tender age of fourteen, Einstein had a firm grasp of integral and differential calculus.

When he was fifteen years old, Einstein dropped out of school because he hated the strict curricula. He resolved to devote the next year to self-study. He took a test to go to college when he was sixteen years old and despite failing in language and history, he did well in the science and math sections. He continued his education in a secondary school that valued his independence of thought and expression.
He went on a self-guided study journey, reading books on many disciplines, driven by his inexorable need for information. His insatiable curiosity led him to pursue knowledge in every field, from philosophy to mathematics.

About the school, Einstein once wrote, "It is, in fact, nothing short of a miracle that the modern methods of instruction have not yet entirely strangled the holy curiosity of inquiry".

His perseverance was rewarded in due course and was admitted to the esteemed Zurich-based Swiss Federal Polytechnic. It was there that he refined his scientific abilities and started to produce discoveries with far-reaching consequences.

After completing his studies, Einstein's genius started to shine brightly. He published several pioneering papers that challenged the existing scientific theories of the time.

But he wasn't just a brilliant scientist; as mentioned he also had a quirky and playful side. He was known for his unruly hair, which became his trademark. Rumor has it that he once said, "I don't need a comb; my hair is a natural antenna for brilliant ideas!"

A. Einstein

Einstein's fame grew exponentially, and he became a household name. People would eagerly await his public lectures, where he would captivate audiences with his wit and charm. He had a knack for explaining complex scientific concepts in simple terms, making them accessible to everyone.

7.2 The Relativity Theory

The relativity theory introduces the concept of spacetime, which combines the three dimensions of space with the dimension of time into a four-dimensional framework. This revolutionary idea challenged the classical notions of absolute space and time and provided a new perspective on the nature of reality.

But let's see how it all began.

Einstein's work was heavily influenced by two of the most iconic physicists of all time. First, there were the laws of motion of Sr. Isaac Newton and second, there were the laws of electromagnetism laid down by James Maxwell.

Newton's law insisted that velocities are never absolute but always relative so their magnitude must be appended by the phrase "with respect to". For instance, a car is traveling at 60 km/h with respect to someone at rest.

However, it only travels 20 km/h with respect to a car traveling 40 km/h in the same direction. This is also true for the velocities of Earth, the Sun, and the entire Milky Way galaxy.

On the other hand, Maxwell found that the speed of an electromagnetic wave, such as light has a fixed value. It travels at the exorbitant speed of 299,792,458 m/s regardless of who observes it. However, Maxwell's notion seems incompatible with Newton's notion of relative velocities. If Newton's laws are truly universal, why should the speed of light be an exception?

This presented Einstein with a disheartening dilemma. Einstein had to make an arduous choice, either Newton's laws were incomplete or the speed of light was not a universal constant.

Einstein realized that the two notions could coexist with a small tweak in Newton's laws. To get rid of the discrepancy in the measurements, Einstein suggested that the time itself for someone in movement must slow down to compensate for the increase in speed such that the magnitude remains constant. Einstein called this absurdity *"Time Dilation"* and his newfound theory *"Special Relativity"*.

How space and time are intertwined

Rather than being static and independent, space and time are related to the movement of the observer. They are two parts of a larger concept called precisely Spacetime.

Envision a toy vehicle racing across an infinite level surface at a constant speed of 100 kilometers per hour.

Now we may mark the north and east directions with arrows on this surface.

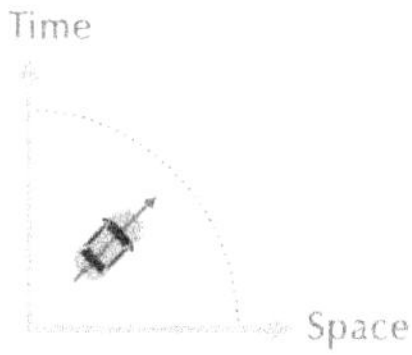

For the sake of argument, let's say that the automobile can only go in three straight lines: east, north, and northeast. Traveling northeast would imply that the car is actually traveling in two directions at once, with equal amounts of velocity in each.

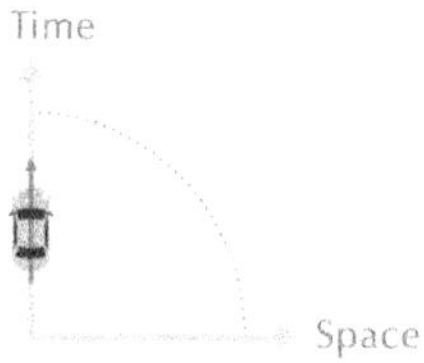

We can therefore mix this idea with our analogy. An object can move vertically, traveling only through time and not moving through space (that's probably what you're doing right now, you're sitting reading this book, so your position in space isn't changing).

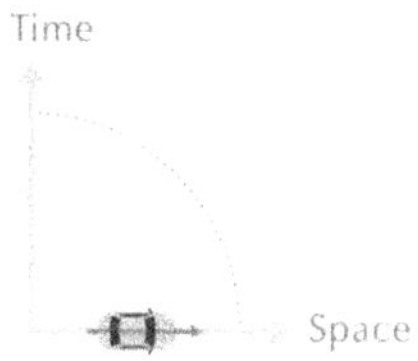

But what happens when you actually begin to go through space? As you start to travel across space, you'll notice that you're moving less through time. You may forget about passing through time altogether if all you do is go through space.

Your clocks will slow down as your speed increases, and they will almost stop as you approach the speed of light, according to *The General Theory of Relativity*.

Newton believed that the time moved unflinchingly in a single direction: forward. Einstein however had just realized that time stretches and contracts varying with velocity. Due to its malleability time like space deserved its own dimension, in fact Einstein claimed that the two were one and the same. Together they formed a four-dimensional fabric or continuum called precisely spacetime.

A brilliant idea that some scientists came up with, was to accelerate some radioactive particles - since we physically cannot travel at the speed of light due to our mass, a series of experiments have been set up to test whether we can really see the malleability of time.

It was taken in observation the Muons decay which is a radioactive process that takes about 6-7 minutes to be completed if left in the wild when it's not part of an atom. The particles were sped up in a particle accelerator at an incredible velocity, and it was observed that the "internal wristwatch" of these particles significantly slowed down making them live longer.

In other words, the "life expectancy" of these accelerated muons was greater than that of muons not accelerated, taking more minutes to decay. Isn't that crazy!? That is a kind of living proof of Einstein's theory.

In this wake, let's now imagine what it would mean to travel at the speed of light. Let's imagine a bunch of photons (which are actually the only particles that can travel at that speed) had a clock, as difficult as it is to picture, that clock would never tick. Which means they do not experience time.

In looking out the window, the light that enters into your eyes, from the Sun and from other stars of the Milky Way, have emitted (probably some of them) their photons more than 30.000 years ago. That light has traveled through the gaps of interstellar space and moved through obstacles of that space, through Earth's atmosphere, then eventually came down and landed in front of your eyes.

These photons, because they've traveled at that speed, have experienced something incredible. These photons when were born, at the star that emitted them, were detected by your lens eye in the same instant! According to the photon perspective. In other words, in traveling only on the space axis, the photons had no knowledge of the trip, because time didn't exist for them!

So just the idea that light can move across the universe and have no age, it's a fascinating fact to most physicists.

7.3 How Mass Influences Spacetime

The fabric of space and time is fundamentally shaped by mass. Fundamental to Einstein's general theory of relativity is the idea of mass and how it interacts with spacetime.

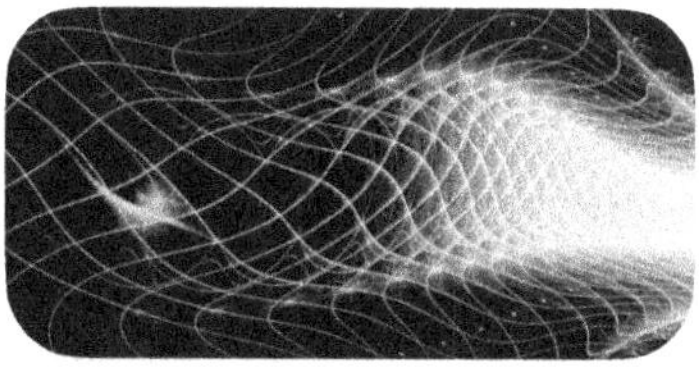

Illustration of the Spacetime fabric

Things that are heavy or take up room in the physical world are often what come to mind when we consider mass.

But in the framework of general relativity, mass transcends its classical meaning and becomes a feature that distorts the universe's fundamental geometry.

Spacetime, as mentioned, is the four-dimensional framework in which all events take place. He revolutionized this field by treating time as if it were almost a geometric parameter and merged it with the other three spatial dimensions (height, width, and depth). Not only that, Einstein believed and then proved that mass causes spacetime to curve, creating what we perceive as gravity.

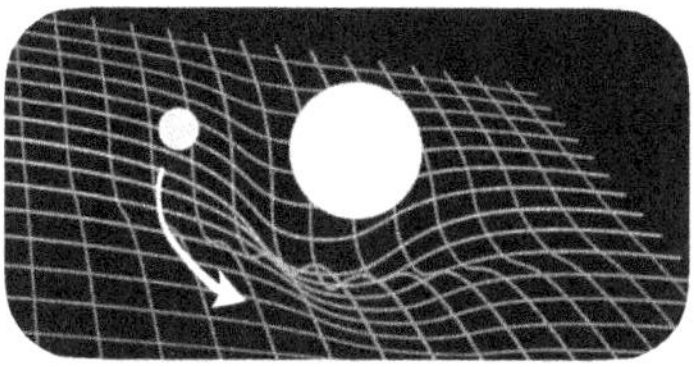

Imagine a massive object placed on a flat, elastic sheet or a trampoline. The object creates a depression in the sheet, causing nearby objects to roll towards it. Similarly, mass bends the fabric of spacetime, causing other objects to move along a curved path towards the mass. This curvature of spacetime is what we experience as gravity.

The more massive an object, the greater its influence on the

curvature of spacetime, like for example, a star with immense mass, or a black hole, creates a significant curvature in spacetime.

Planets, such as Earth, Mars, and Jupiter, are subject to the gravitational pull exerted by the sun's curved spacetime.

This gravitational force compels them to follow specific paths, known as orbits, around the sun.

Even smaller objects within the solar system, like asteroids, experience the effects of spacetime curvature. As they traverse through space, their trajectories are influenced by the curved spacetime created by the sun's immense mass. This influence determines their paths, causing them to revolve around the sun or even interact with other celestial bodies in intricate ways.

It is truly remarkable to consider that even humans, with our comparatively minuscule mass, are impacted by the curvature of spacetime. Our very existence and movements on Earth are governed by the gravitational field created by the sun's spacetime curvature. It is this force that keeps our planet in a stable orbit and allows us to experience the rhythms of day and night.

One fascinating consequence of this interplay between mass and spacetime is the phenomena of gravitational waves. When massive objects, such as two black holes, orbit each other, they create ripples in the fabric of spacetime, propagating outward like waves. These gravitational waves carry valuable information about the objects and events that generated them, providing scientists with a new way to explore the cosmos.

So, in short, mass is not just a property of objects but a fundamental factor that shapes the very nature of the universe.

It influences the curvature of spacetime, giving rise to the force of gravity that governs the motion of celestial bodies. In a nutshell: matter tells spacetime how to curve, and spacetime tells matter how to move.

Sheet or Lattice?

Viewing spacetime as a sheet allows us to visualize it as a two-dimensional surface that can be curved or distorted by massive objects. Similarly, in this sheet-like representation, the presence of massive objects creates curves and warps in spacetime, influencing the motion of other objects around them. This representation is more of a simplification for scholarly purposes and considering it as a 2D surface helps us in understanding the concept of gravitational waves.

Just as ripples propagate across the surface of a pond when a stone is thrown in, gravitational waves are the ripples in the fabric of spacetime, generated by the acceleration of massive objects. This analogy provides a tangible way to grasp the propagation of these waves through the universe.

The lattice representation

Alternatively, conceiving spacetime as a grid offers a different perspective. In this analogy, spacetime is represented as a three-dimensional lattice, with each point on the grid denoting a unique position in spacetime. This grid-like representation allows for a more precise calculation of distances.

Considering spacetime as a grid also enables us to comprehend the concept of spacetime intervals, which describe the separation between two events. By assigning coordinates to each point on the grid, the mathematical framework of relativity becomes more accessible, facilitating calculations and predictions relating to the nature of spacetime.

Since it's a 4D structure, the reality is much more similar to this last description, but ultimately, whether one imagines spacetime as a sheet or a grid, both analogies offer valuable insights into the nature of this fundamental construct.

The sheet analogy provides a visual understanding of how massive objects curve and warp spacetime, while the grid analogy allows for precise calculations and predictions within the framework of relativity.

7.4 Connection between mass and energy

Einstein's equation, **$E=mc^2$** is probably the most "pop" equation of all time and has without any doubt expanded our understanding forever - in its simplicity, it tells us something very deep, that mass is equivalent to energy, and that one can be converted to the other and vice versa.

Conceptually it doesn't quite make sense, how something ethereal like electricity or heat can be the same as something tangible like a dumbbell or a fat cat?

This cornerstone of modern physics reveals a profound relationship between mass and energy, demonstrating that they are not separate entities but rather different forms of the same fundamental essence. This revolutionary insight has had far-reaching implications, reshaping our understanding of the universe and paving the way for numerous technological advancements.

Now let's break down this formula together.

Let's start with the mass '**m**', in classic physics, mass is simply a measure of the amount of matter that makes up an object, in other words, how much "stuff" there is in a specific object - the more an object is massive, the more matter it contains.

Mass is often confused with the weight of the object, which is completely another concept. Weight is the force with which mass is attracted by our planet. If we used a scale on the Moon and weighed an object, the total weight of this object would be less heavy compared to the Earth, despite having the same mass.

However mass manifests another important characteristic that Newton observed and understood first, and that is that it's very difficult to accelerate an object. The force needed to accelerate an object for example from 0 to 100km/h is proportional to the mass of the object. If I have to accelerate an object that has double the mass of another, I will need to double the force. Or if I apply the same force to two objects in which one has double the mass of the other, the one with greater mass will accelerate by half. In telling this Newton states that the mass of an object is a measure of how much an object resists the change of state, in this case, the velocity. This is also called **inertia**.

Now, let's continue with '**E**' the energy. This is a quantity that is a little more complicated to grasp intuitively. There are various forms of energy (potential, kinetic, thermal, mechanical, chemical, etc.) but what all of them have in common is that they can be transferred and used to do a job, or to complete a task.

The peculiarity of energy is that it can transform from one form to another under certain conditions, for example, the combination of potential energy and kinetic energy that occurs when you are on a roller coaster. Energy cannot be created nor destroyed, it can only change its form and in a closed system, the total value of that energy remains constant.

Until Einstein, no one had ever directly related the mass and the energy of an object because until that moment they were two completely distinct entities. But in 1905, Einstein discovered that an object simply by having a mass possesses energy - and this energy is in relation to the mass of that object multiplied by the speed of light squared represented by '**c**'.

It's important to point out that this is true for an object completely at rest and that therefore has no kinetic energy, it is also true for an object that doesn't even have potential energy because far from gravitational fields, and true for an object that doesn't have any other form of energy stored in it, such as chemical energy or heat. So, according to this equation, any object with a mass has always its energy multiplied by $\mathbf{c}^2$. Mass and energy are two distinct ways of seeing the same thing and cannot be separated.

As we said before, the speed of light reaches almost 300,000km/s, now take that value and square it, then multiply it by the mass of the object and you will realize that the total energy is immense! To give you an idea, the energy trapped in a single gram of matter would be enough to keep a 100-watt light bulb lit for 30 thousand years!

Obtaining Energy from Mass

The simplest (but also the most primitive) way to extract energy from mass is combustion. As said, everything that exists (molecules included) is in some form of energy state. In burning the molecules, let's say of a log, we're extracting from its fibers heat and light. It's due to the chemical potential energy in the molecules.

Where did that energy come from? Very simple, the log used to be a tree, and the tree got its energy from the sun! The sun built those molecules that contain stored energy, so that's why wood burns!

Another example is calories! The ones you get from food. It's always chemical energy from living beings (plants and animals), once consumed, they provide all you need for your body to function perfectly. If you consume more calories than you need, your body says: let's store them! And create chemical potential energy in the form of fat. And it will remain preserved there waiting to be burned, as similarly happens to a pile of logs.

So, next time someone tells you you're fat, just reply: "I'm not fat, I'm just filled with potential!"

You see? Our existence and everything we do is nothing but a ballet of conversion of energy from one form into another. Another process (this time more sophisticated) of obtaining energy from mass is exemplified in nuclear reactions, such as nuclear fission and nuclear fusion.

Nuclear fission involves splitting the nucleus of an atom, resulting in the release of a tremendous amount of energy. This process was famously harnessed in the development of the atomic bomb during World War II. Today, nuclear fission is primarily utilized in nuclear power plants to generate electricity.

On the other hand, nuclear fusion involves combining the nuclei of atoms to form a heavier nucleus. This process occurs at extremely high temperatures and pressures, such as those found in the core of the Sun. Nuclear fusion has the potential to provide a nearly limitless source of clean and sustainable energy, although it remains a significant technological challenge to achieve controlled fusion reactions.

Obtaining Mass from Energy

The reverse process, obtaining mass from energy, is less commonly observed in everyday life but is nonetheless equally significant. While it is not feasible to directly convert energy into mass in a controlled manner, it is possible to observe this phenomenon in high-energy particle physics experiments.

In particle accelerators, such as the Large Hadron Collider (LHC), particles are accelerated to extremely high speeds and collide with each other. These collisions produce a tremendous amount of energy, which can momentarily condense into massive particles.

These particles, often short-lived, provide valuable insights into the fundamental building blocks of matter.

The consequences of converting energy into mass go well beyond the field of particle physics. By facilitating research into the characteristics and interactions of particles, they help advance disciplines like particle physics and quantum mechanics, which in turn enhance our comprehension of the cosmos at its most basic level.

7.5 E=mc² is incomplete

So far we have learned that this formula accurately describes what really happens in numerous phenomena in our daily life and that the energy and the mass are really two faces of the same coin.

But what we have considered so far with this equation is only the mass of objects that are stationary. What about objects in movement?

According to Einstein's theory of relativity, as an object accelerates and approaches the speed of light, time dilation occurs. This means that time appears to slow down for the moving object relative to a stationary observer.

He introduced what is called the Lorentz Factor, which is a term used in physics to describe the relativistic effects on time, length, and mass when an object is moving at speeds close to the speed of light. The Lorentz Factor, denoted by the symbol γ (gamma), quantifies this time dilation effect.

The Lorentz Factor is defined as:

$$\gamma = \frac{1}{\sqrt{1 - \frac{v^2}{c^2}}}$$

where:

- v = is the relative velocity between the observer and the moving object.
- c = is the speed of light in a vacuum.

Components Explained

Relative Velocity (v):

- This is the speed at which the object is moving relative to the observer. It's a crucial variable as the Lorentz Factor increases significantly as **v** approaches **c**.

Speed of Light (c)

- As we learned, it is a fundamental constant in the universe, **c** is approximately 3×10^8 meters per second. It represents the maximum speed at which all energy, matter, and information in the universe can travel.

Squared Root Term $\sqrt{1 - \frac{v^2}{c^2}}$

- This term approaches zero as **v** approaches **c**, which causes the Lorentz Factor (**γ**) to increase towards infinity. This reflects the dramatic effects on time, length, and mass as objects move at speeds close to the speed of light.

As the velocity of an object approaches the speed of light, the Lorentz Factor becomes increasingly larger. When the object is at rest (v = 0), the Lorentz Factor is equal to 1, indicating that time dilation is not present. However, as the velocity increases, the Lorentz Factor diverges towards infinity.

Therefore an object close to the speed of light has a Lorentz Factor with an infinite value, in fact, to accelerate an object to the speed of light we would have to spend infinite energy and this is one of the reasons why the speed of light is an insurmountable limit.

Another way of looking at this concept is to think that when the energy of an object increases, its inertia also increases, that is, it becomes increasingly difficult to accelerate it.

A question to ask would also be: does the mass of an object change if it moves with respect to us? If energy is equivalent to mass, and that object travels at an incredible velocity, it will have an incredible mass too?

Not, really. Physicists theorize that eventually exists a single and only mass, and it is the mass measured when the object is stationary with respect to us. This concept, proposed by physicists, suggests that there is a fundamental mass that remains constant regardless of the object's motion. However, while Newton's understanding of inertia was solely based on the object's mass, Albert Einstein introduced the revolutionary idea that energy also contributes to inertia.

So, as crazy as this idea may seem, when we administer energy to an object, some of that energy will accelerate it, increasing its kinetic energy and some of it will increase its inertia. The existence of increased inertia is a fundamental factor that prevents most bodies in the universe from reaching the speed of light.

Another important characteristic to know regarding mass is its total value when it's composed of different elements. If we take a system that is stationary with respect to us but made up of parts that are in movement, we can observe that it has a mass that is greater than the sum of its components.

In the case of protons, for example, they are made up of quarks, which are particles that are in continuous agitation and extremely bonded to each other. The total mass of the proton is greater than the sum of the masses of the individual quarks that compose it - this is because the kinetic energy produced by the movement of the quarks contributes in part to the total mass of the system. Stupefying, right?

7.6 Why we can't go faster than the speed of light?

Let's see in this section the reasons why experimentally and theoretically is impossible to travel faster than the speed of light and let's summarize the concepts illustrated so far in this chapter.

As we have seen, Einstein's theory states that as an object accelerates towards the speed of light, its mass increases, and the amount of energy required to continue accelerating it also increases. This phenomenon is known as *relativistic mass increase*.

It would potentially take an endless quantity of energy to get the object to or beyond the speed of light since its mass would grow infinite as it gets closer to that speed. And as you may conclude, it is just not possible to meet such an endless energy need.

Time Dilation

Another consequence of the Special Theory of Relativity is time dilation. As an object approaches the speed of light, time slows down for that object relative to a stationary observer. This means that time passes more slowly for the moving object compared to an observer at rest.

If an object were to accelerate beyond the speed of light, time dilation would become even more extreme. Time could potentially run backward for the object, which violates the fundamental causality principle in physics. Cause and effect relationships would become jumbled, leading to logical inconsistencies.

Mass-Energy Equivalence

Einstein's famous equation, $E=mc^2$ shows that mass can be converted into energy, and vice versa. As an object accelerates towards the speed of light, its kinetic energy increases significantly. Approaching the speed of light would require an infinite amount of energy, and this would imply an infinite mass-energy equivalence.

The concept of infinite energy or infinite mass violates the laws of physics as we currently understand them.

Consequences for the Universe

If it were possible to exceed the speed of light, it would have profound consequences for our understanding of the universe. The principles of causality, time, and energy conservation would be fundamentally challenged. The fabric of space-time itself would be disrupted, leading to paradoxes and inconsistencies.

Moreover, the speed of light is intimately connected to the behavior of electromagnetic waves, which include visible light, radio waves, and X-rays, among others. It is through the study of these waves that we have gained many insights into the nature of the universe. If the speed of light could be surpassed, it would disrupt the entire framework of electromagnetic theory.

Let's have fun for a moment with something that FOR NOW is considered pure fantasy, let's talk about ***Tachyons***.

You may have come across this word if you are a fan of science fiction comic books - the term Tachyon, which comes from the Greek word *"tachys"*, which means fast, was coined by the late Gerald Feinberg of Columbia University.

According to physical theory, tachyons are hypothetical particles that travel faster than light and have "imaginary" masses, they're merely hypothetical since they break several established physics laws and their existence indicates that the theory that contains them is unstable.

Normally, events are determined solely by what has occurred in the past. However, features of Einstein's theory suggest that a tachyon could be identified before it formed. Because the presence of a tachyon would contradict Einstein's extraordinarily successful theory, a claim of tachyon detection would necessitate extraordinary evidence.

The theory states that nothing can travel faster than light, to be precise nothing with a mass greater than zero can travel to the speed of light (or higher) because, as we've seen, this would demand infinite energy and result in infinite mass. Only massless particles such as photons and gluons, can travel at the speed of light - as a result, according to the relativity theory, a particle will not be able to accelerate faster than that limit.

But what if it was always that fast, from the beginning of time? What if it was created in this manner during the Big Bang?

If such a particle existed, it would be unable to slow down to the slower speed of light of today, it would have the strange property of requiring less energy as its velocity increased, hence approaching its lowest energy state as it reached infinity's speed.

To slow down to the speed of light, on the other hand, would require infinite energy, and hence would never be able to cross that speed barrier. As a result, particles traveling at different speeds on either side of the speed of light encounter a two-way barrier.

That would not violate the law of relativity, but it would have some strange properties. One peculiar property is that tachyons would have an imaginary rest mass, the other strange property is that depending on the relative arrangements between the frames and spacetime, an observer could see tachyons travel the other way through time, thereby seeing them before they're produced (in the observer's frame).

So, while tachyons represent a fascinating possibility, their existence, as said, remains speculative. The laws of physics as we currently understand them prevent anything with mass from reaching or exceeding the speed of light. However, scientific exploration and the quest for knowledge continue to push the boundaries of our understanding, leaving room for exciting discoveries and breakthroughs in the future.

Fun fact: There exists only one way we can actually travel faster than light on Earth. Here's the trick.

We can travel faster than light in a medium where light travels at a lower speed than it would in a vacuum. Like water, for example! The same happens when light goes through glass, and when it goes through diamonds. Just to picture you more clearly the concept: in a diamond, light travels 40% as fast as it does in empty space.

And you know what? We are able to make travel particles faster than light in those mediums! Wow! But what exactly happens when we do that? It was tested and we found out that when we make that happens, there's the manifestation of a flash of light. It's called the Cherenkov Radiation.

Cherenkov Radiation is a fascinating phenomenon that occurs when a charged particle passes through a dielectric medium at a speed greater than the phase velocity of light in that medium. It was first discovered by the Soviet physicist Pavel Alekseyevich Cherenkov in 1934. When a charged particle, such as an electron or a proton, moves through a dielectric medium, it polarizes the surrounding atoms or molecules. This polarization disturbs the electromagnetic field in the medium, causing the emission of electromagnetic radiation.

The Cherenkov Radiation appears as a characteristic blue glow or sometimes as ultraviolet light. The color of the radiation depends on the properties of the dielectric medium through which the charged particle is passing. The intensity and direction of the Cherenkov Radiation are dependent on the velocity and charge of the particle, as well as the refractive index of the medium. The radiation is emitted in a cone-shaped shockwave, with the apex at the point where the charged particle enters the medium.

We can find an analogy with the sonic boom, it's conceptually similar - when a fighter plane goes faster than the speed of sound in the medium then there's this shock wave that comes out upon doing so.

QUANTUM FIELD THEORY

Before delving into the intricacies of Quantum Field Theory, it is essential to understand the historical and theoretical background that led to its development.

Quantum mechanics revolutionized our understanding of the microscopic world by introducing the concept of wave-particle duality and probabilistic interpretation. However, it posed challenges when dealing with relativistic phenomena, where the principles of special relativity had to be taken into account.

Albert Einstein's theory of special relativity provided a framework for understanding the behavior of objects moving at relativistic speeds. It established that the laws of physics should remain invariant under Lorentz transformations, which led to a new understanding of space and time.

At its core, QFT is based on the concept of a quantum field, which permeates all of space and time. These fields are not fixed entities but rather dynamic quantities that can fluctuate and interact with particles. In QFT, particles are considered as quantized excitations of their respective fields.

The principles of QFT are encapsulated in a mathematical framework known as the Lagrangian formalism. The Lagrangian formalism describes the dynamics and interactions of fields and particles through a set of equations called the field equations of motion. These equations govern the behavior of the fields and determine the probabilities of different particle interactions.

Quantum Field Theory has far-reaching applications in various branches of physics, including particle physics, condensed matter physics, and cosmology. But let's see more in detail.

8.1 The Limits of Quantum Mechanics

As we have already learned, unlike any object that has a definite position at the macroscopic scale, a particle doesn't have a clear position, its presence is distributed throughout space with more or less probability. This description of particles as waves of probability is called quantum mechanics.

While quantum mechanics does a good job of describing how a particle changes over time, there are a few things that it doesn't account for.

To start, as the quantity of particles changes over time, quantum mechanics breaks down. On the other hand, particles may manifest and vanish in nature all the time, like when an electron in an atom absorbs a photon.

Secondly, quantum mechanics treats each particle independently. In the family of particles such as the electrons, all exhibit the same properties and behaviors. How is it that an electron coming from the far reaches of the cosmos, has exactly the same mass of charge as an electron in an apple?

To answer this question, it was necessary to build a new mathematical framework, a more general description that reconciles quantum mechanics with special relativity.

The Schrödinger equation works quite well in describing the wave nature in quantum objects, but it has some limits. It doesn't work for special relativity and this happens specifically when velocity is close to the speed of light. To account for relativistic effects, space and time must be treated equally (since part of the same grid). But if we look at the Schrödinger equation, we will see that time and space are not treated equally. So it's not that obvious how to treat all the ingredients.

In the 1930's scientists like Paul Dirac realized that one could get an even better formulation of quantum mechanics than Schrödinger achieved, by considering quantum objects in terms of quantized fields.

When this is done, the equations of quantum mechanics could be modified to account for special relativity by treating space and time equally. This led to Quantum Field Theory.

Conformation of the quantum fields

In QFT what we call particles are depicted as layers of quantized waves, in a field that spreads throughout spacetime. The idea of quantized fields was a necessary step in order to describe speedy objects like photons and others that demand to be treated differently, according to the rules of special relativity, when they start to have significant effects at, or close to the speed of light.

If you picture in your mind a calm sea, you will have analogously an image of a quantum field (without any particles). This is our starting point. It may sound a little bit bizarre but imagine that this field or sea stretches throughout all of spacetime (note that this would be three dimensions, but for visualization purposes, we're representing this in two dimensions like the surface of an ocean).

Now, let's make emerge a little wave in the sea, this will correspond (in the vision of this theory) to a particle in our field, it can exist multiple if not infinite particles in the same field or infinite particles upon them, so multiple layers.

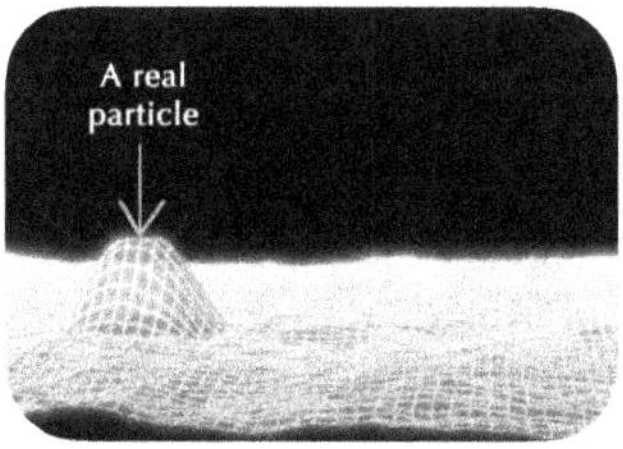

Here each particle is an excitation or a disturbance of the field itself, just like a wave in the sea. Because we're considering our particles as waves, if collided, then the particles collide like waves just as in the double slit experiment. There's however at least one difference between the quantum field and our analogy: our sea is not quantum. For our sea to be quantum, we have two requirements:

- First, the waves in our sea must have some discrete magnitude, because in quantum mechanics things have discrete quantizations and this can be represented by the amplitude or height of the wave. Let's suppose that the waves can only be 1m, 2m, 3m, and so on. Each meter corresponds to the number of particles in any one place.
- Secondly, a real quantum field is never calm. It always has some minimum energy state. So we will say that the 1m waves represent this minimum energy state of the field.

This minimum energy of 1m is called the vacuum energy. It means that for a real quantum field "empty" doesn't mean nothing, there's still something - some minimum energy is always present. Then each extra meter of amplitude or height is a real particle.

Now the interesting thing about quantum waves is that we can only create taller waves in increments of whole meters. So if we don't have enough energy to add at least one meter to the wave, then we won't have any change to the waves at all. In other words, there can no exist 1.25m, 1.6m, or even 1.9m waves, only 2m, 3m, 5 m, etc.

So our quantum field is full of this minimum 1m waves, and each extra meter is then a real particle. But we can only create these taller waves if we have enough energy to reach the next meter. If there's not enough energy to reach the next meter, no taller wave is created.

Another thing: these quantum waves slosh around and, in some places, there might be momentarily enough energy to create a large 2m wave or particle. But then almost just as quickly as it's formed, the particle vanishes again in the sea of fluctuations. These are analogous to virtual particles that come in and out of existence all around us but are undetectable because they last for too short a time.

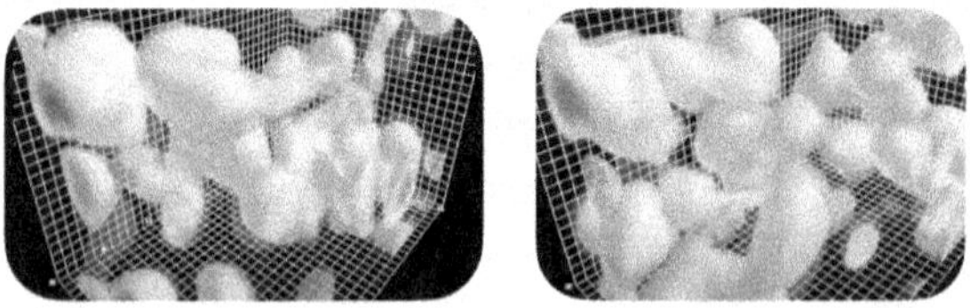

This is a computer simulation of empty space. All the bubbles that you see, are bubbles that are popping and bursting and are perturbations of empy space. Particles are constantly being created and destroyed in this emptiness. But these particles are really excitation of fields that pervade all of space as said.

So in QFT, we have a field that is never still, never empty and we can create a particle if we have the exact amount of energy needed. But still, you should keep in mind that this is just a mathematical construct, that appears to fit reality.

If you expand this concept to other particles we have to imagine other seas on top of seas representing a different field for each fundamental particle that we know of. So there will be a sea representing the field of electrons, for example, a field for photons, a field for quarks, etc.

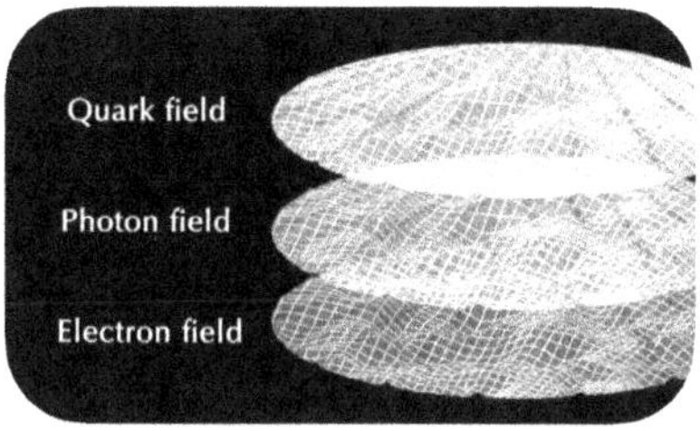

This concept led us to where we are now, we have this general theory called the Standard Model in which we include all known fundamental particles, and we assign to each type of particle its corresponding field. Under this new vision, we can say that every fundamental particle is an excitation in a quantum field that is in constant flux. There are at least as many fields as there are particles in the standard model.

Each particle can propagate in its field and the interaction of these fields can generate an exchange of energy resulting in an endless process of creation and annihilation of particles.

Now you may ask, are fields a real thing, or are they just mathematical constructs? Even though a field (whether portrayed in 2D or 3D) has no known substance, physicists regard fields to be genuine physical stuff. This is because they exist in space and possess energy. Furthermore, their properties can be estimated and correctly predicted using experimental data.

8.2 Standard Model of Particles

Before continuing with quantum fields, we must take a moment on some particles that have been discovered to be the components of what we have until now considered to be indivisible such as protons and neutrons and their associated forces.

The Standard Model of Particles is a general theory that explains and classifies all the real fundamental and indivisible particles in quantum physics. It provides a detailed explanation of these particles, including their discovery and how they were obtained.

The Standard Model encompasses three of the four fundamental forces in nature, namely:

- the weak nuclear force;
- the strong nuclear force;
- the electromagnetic force;
- It also includes all the known elementary particles, such as quarks, leptons, and gauge bosons.

Firstly, let's see the nuclear forces that we haven't talked about yet:

The Weak Nuclear Force

The weak nuclear force, also known as the weak interaction, is one of the fundamental forces of nature. It is responsible for certain types of radioactive decay and is involved in processes such as beta decay, where a neutron transforms into a proton, emitting an electron and an antineutrino.

The weak nuclear force is mediated by particles called **W** and **Z** bosons. These bosons are responsible for the transfer of the weak force between particles. The weak force is unique among the fundamental forces as it can change one type of elementary particle into another. This phenomenon is known as flavor change, where a particle can transform into a different particle of the same generation.

One of the significant implications of the weak nuclear force is its role in the synthesis of heavier elements during stellar nucleosynthesis. In the core of stars, where extremely high temperatures and pressures exist, fusion reactions occur, facilitated by the weak force, leading to the production of elements beyond hydrogen and helium.

The Strong Nuclear Force

The strong nuclear force, also known as the strong interaction, is another fundamental force that binds atomic nuclei together. It is responsible for holding protons and neutrons together within the nucleus, overcoming the electromagnetic repulsion between protons.

The strong nuclear force is mediated by particles called gluons. Gluons carry the strong force between quarks, which are the elementary particles that make up protons and neutrons. The strong force is unique in that it gets stronger as particles move apart, unlike the other fundamental forces.

The strong nuclear force is responsible for the stability of atomic nuclei. Without the strong force, the electromagnetic repulsion between protons would cause nuclei to disintegrate. It also plays a crucial role in nuclear reactions, such as nuclear fusion, where light atomic nuclei combine to form heavier ones, releasing vast amounts of energy.

Now let's see these sub-sub atomic particles one by one.

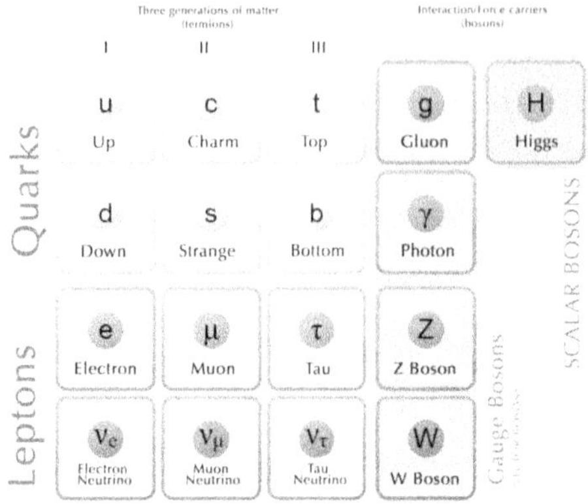

Quarks.

Quarks are a group of building blocks of protons and neutrons, which are the particles that make up the atomic nucleus. There are six types of quarks: up, down, charm, strange, top, and bottom. Each quark carries a fractional electric charge and has a corresponding antiquark with an opposite charge.

Leptons.

Leptons are a family group of elementary particles that do not participate in the strong nuclear force. They include electrons, muons, and tau particles, along with their corresponding neutrinos.

Muons: The Heavy Siblings of Electrons

Muons, often referred to as "heavy electrons," belong to the class of particles called leptons. They bear an electric charge, albeit with a magnitude identical to that of an electron, but have a mass around 200 times greater. Muons are produced naturally through high-energy processes such as cosmic ray interactions in Earth's atmosphere or artificially through particle accelerators.

One might wonder why muons are of such interest to physicists. Well, their behavior and characteristics offer valuable insights into the fundamental forces and particles that govern the universe. Muons, being unstable, decay into other particles, allowing scientists to study the laws of particle interactions and delve into the mysteries of particle physics.

Taus: The Elusive and Short-lived Particles

Taus, similar to muons, are also classified as leptons. They possess significant mass, around 3,500 times that of an electron, and carry an electric charge. However, unlike muons, taus have an extremely short lifespan, decaying rapidly into other particles. Due to their fleeting existence, studying taus presents a unique challenge to scientists.

Despite their short lifespan, taus play a crucial role in the ongoing quest to understand the fundamental nature of matter and energy. Their study provides insights into the weak nuclear force, one of the four fundamental forces of nature, which governs processes such as radioactive decay and nuclear reactions.

Neutrinos: The Ghostly Messengers of the Universe

Neutrinos are enigmatic subatomic particles that have fascinated scientists for decades. Despite their elusive nature, these ghostly messengers play a crucial role in our subatomic universe.

Unlike other particles, neutrinos have a minuscule mass, making them incredibly difficult to detect, and are electrically neutral, which means they do not interact with electromagnetic fields. This property allows them to effortlessly pass through matter, earning them the nickname "ghost particles."

First proposed by physicist Wolfgang Pauli in 1930, neutrinos were initially met with skepticism. However, subsequent experiments confirmed their existence, and scientists soon realized their significance. Neutrinos are produced in various processes, such as nuclear reactions within stars, radioactive decay, and high-energy collisions in particle accelerators.

Furthermore, neutrinos have the unique ability to travel vast distances without being significantly affected by intervening matter or magnetic fields. This makes them valuable tools for studying distant cosmic phenomena, such as active galactic nuclei and gamma-ray bursts.

Neutrino detectors, such as the Super-Kamiokande in Japan and the IceCube Neutrino Observatory in Antarctica, have been instrumental in unraveling the mysteries of these elusive particles. These detectors consist of massive tanks filled with ultra-pure water or ice and are equipped with sensitive photodetectors to capture the faint flashes of light produced when neutrinos interact with the surrounding medium.

Bosons.

Bosons are another group of elementary particles that are important in the field of particle physics. They are named after the Indian physicist Satyendra Nath Bose, who made significant contributions to the foundation of quantum statistics. They are characterized by their integer spin, which distinguishes them from fermions, particles with half-integer spin.

There are various types of bosons, each with its unique properties and significance.

- Photon: Photons are the most famous bosons, one of their job is to facilitate electromagnetic interactions. They are responsible for carrying the electromagnetic force, including light and other forms of electromagnetic radiation.

- W and Z Bosons: W and Z bosons are particles that mediate the weak nuclear force. They play a crucial role in processes such as radioactive decay and are responsible for the phenomenon of electroweak unification.

- Gluon: As mentioned, Gluons are bosons that mediate the strong nuclear force, which is responsible for holding atomic nuclei together. They are crucial in understanding the behavior and structure of protons, neutrons, and other particles composed of quarks.

- Higgs Boson: The Higgs Boson is a fascinating particle that has captivated the attention of scientists and researchers around the world. This elusive particle holds the key to unlocking mysteries about the fundamental building blocks of reality. Scientists discovered that the Higgs Boson plays a vital role in the mechanism of mass generation and it interacts with other particles, such as quarks and leptons, through the Higgs field, imparting mass to them. This interaction is crucial for the formation of atoms, molecules, and ultimately, all matter in the universe. Let's look at it better in the following section.

8.3 The Genesis of Matter

The discovery of the Higgs boson, sometimes called the "God particle," occurred in 2012 at the Large Hadron Collider (LHC) in Geneva. It's considered the "*most basic particle*" in physics because we know more about the cosmos than ever before thanks to this groundbreaking discovery.

The Higgs boson (and its homonym field) is essential because it possesses a very unique and peculiar property that distinguishes it from other quantum fields.

It's able to give birth to mass in interacting with other fields, as stated in the Standard Model of Particle Physics.

It is named after the physicist Peter Higgs, who proposed its existence and was awarded the Nobel Prize in Physics in 2013 for his contributions to the discovery of this intriguing particle.

Peter Higgs

Peter Higgs, along with other physicists such as Robert Brout, François Englert, and others, independently proposed the existence of this strange particle/field in the 1960s. Their work was based on theoretical frameworks developed by earlier physicists, including Philip Anderson and Yoichiro Nambu.

The Higgs field is characterized by its energy distribution throughout space. It is through the interaction with this field that particles acquire their mass. Imagine particles as tiny entities swimming through a sea of energy. As they navigate through the Higgs field, they experience a drag-like effect, akin to swimming against a current.

Interestingly, not all particles interact with the Higgs field in the same way. Some particles, like the W and Z bosons, interact more strongly and thus acquire more mass. Other particles, such as photons, experience little to no interaction with the Higgs field, resulting in their lack of mass.

Now, because these are quantized fields, any excitation occurs only in predetermined quantities; for example, an excitation of the electron field must occur in integer multiples of 0.511 mega electron volts (MeV), which is an electron's mass. So the field's energy can be 1.22 MeV or any other multiple, but not 0.7 MeV for example, which is not an integer multiple of 0.511.

Now, I'd want to point out that the electron's mass is only possible because of its interaction with the Higgs field; without this interaction, an electron would be massless, with energy only in the form of momentum, similar to photons. This massless electron would behave like a charged photon, traveling at the speed of light.

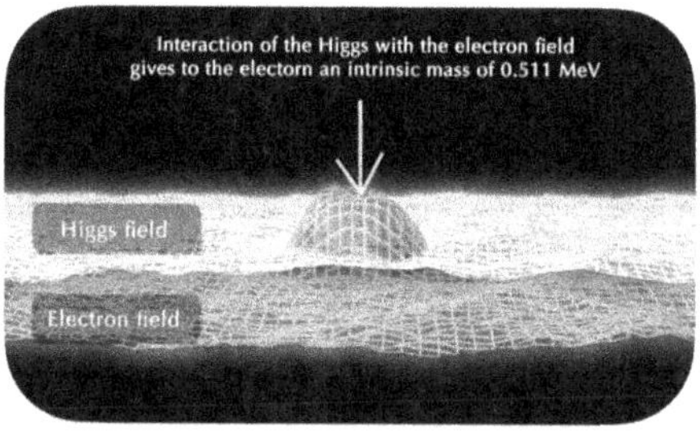

In fact, without the Higgs field, all of the standard model's fundamental particles, with the probable exception of neutrinos, would have no mass.

So the question is how does this mass come about? To comprehend this, we must first grasp the idea of Vacuum Expectation Value (VEV) for diverse fields. What exactly does this mean?

Let's pretend for a moment that there is no Higgs field; if we took any of the fields of any particle and placed them inside an imaginary empty box, and then weighed that box, we would discover that they have no weight. In other words, the fields would have no mass other than quantum fluctuations, despite the fact that virtual particles from those fields would be present throughout.

But here is where the magic kicks in, there's an exception to this rule: the Higgs field. It's unique because this field in empty space, unlike any other field, **has a net positive mass**. Its mass is NOT zero in empty space.

If we weighed the box with the Higgs field inside, we would get a weight. The Higgs field in empty space has mass, which is known as vacuum energy. As previously stated, the vacuum energy is not zero, but rather 246 giga electron volts (GeV), which is the value of the Higgs field when it is in the vacuum state or in its lowest energy state.

Now, when any massless field interacts with the Higgs field's pervasive and intrinsic 246 GeV, it obtains energy, and because energy and mass are equivalent, it also gains mass.

So what we can observe is that the electron is coupled to a field with positive value everywhere, individual electrons are constantly interacting with the Higgs field permanently, which is why an electron from the far reaches of the universe has the same mass and properties as an electron in an apple here on Earth.

These interactions are represented by mathematical equations that encapsulate the exchange of energy and the resulting mass generation. Through these equations, physicists have been able to explain why some particles are more massive than others.

Another very important fact to remark is that the Higgs field is thought to have played a pivotal role in the very early moments after the Big Bang, giving rise to the formation of elementary particles and the structure of the universe as we know it today.

At the genesis of the universe, just moments after the immense explosion, the universe was a hot, dense, and rapidly expanding soup of particles. During this early stage, the fundamental forces of nature were unified and the laws of physics as we know them today did not yet apply.

By imparting mass to other particles, the Higgs boson played a crucial function in shaping the structure of the universe. Without it, particles would have remained massless, resulting in a vastly different cosmos. The presence of mass allows particles to interact with one another through gravitational forces, forming larger celestial bodies such as stars, galaxies, and ultimately, the complex structures necessary for life to emerge.

While the precise details of how the Higgs boson contributed to the genesis of the universe are still the subject of ongoing research, its existence provides a crucial piece of the puzzle in understanding the fundamental nature of our reality. Scientists continue to explore the properties and implications of the Higgs boson, pushing the boundaries of our knowledge.

As for the question of whether it is possible to create matter from nothing, the discovery of this "God Particle" does not directly answer this. While scientists have not yet been able to create matter from nothing, the discovery of the Higgs boson has opened up new avenues of research and deepened our understanding of the fundamental building blocks of the universe.

Similarly, the creation of new matter, or elements or gold from nothing (in case any of you were wondering) is not currently possible, and the discovery of the Higgs boson does not change that fact. Sorry!

The creation of elements, such as gold, requires specific nuclear reactions that are not within our current technological capabilities.

8.4 Feynman Diagrams

When it comes to theoretical physics, Feynman diagrams are invaluable for both visualizing and computing particle interactions. These diagrams, created by scientist Richard Feynman, who won the Nobel Prize in physics, show how mathematical expressions explain the behavior of particles.

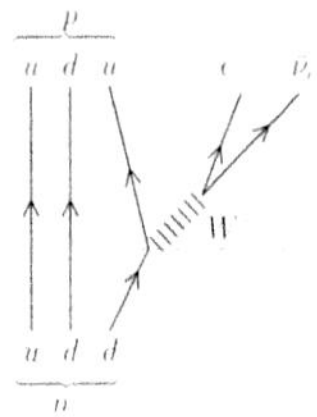

Visualizing the mathematical principles underlying quantum field theory is the essence of Feynman Diagrams. They use lines to represent the particles participating in an interaction and arrows to show which way the particles are moving. Particles collide or interact at the points where the lines intersect.

Examples of diagrams

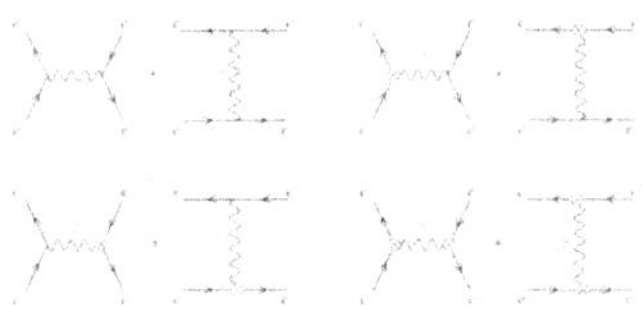

The capacity to simplify complicated computations is one of the main advantages of Feynman Diagrams. Physicists are able to graphically evaluate and calculate the odds of possible events by breaking down particle interactions into graphical pieces. Particles' behavior may be understood and predicted by physicists using these diagrams in a variety of contexts, from the most basic to the most complex.

Feynman Diagrams are fantastic because they may be used in any area of physics. Their applications include the fields of electroweak theory, quantum chromodynamics (QCD), and quantum electrodynamics (QED). Particle decay, dispersion, creation, and annihilation are just some of the events that physicists examine using Feynman Diagrams.

Additionally, Feynman Diagrams mediate between the realms of theory and experiment. By comparing experimental results with predictions made by Feynman Diagrams, physicists may either confirm or improve their theories. The knowledge of the basic forces and particles that control the cosmos has been greatly advanced by this repeated process of theory-experiment interaction.

Although Feynman Diagrams depict the interactions between particles graphically, they should not be taken as depictions of actual physical processes. Their actual purpose is to facilitate computations and the conceptualization of the underlying physics using mathematical means.

8.5 Quantum Electrodynamics

At its core, QED describes the interaction between electrically charged particles and electromagnetic fields. It explains how electrons and photons, which are the fundamental particles of matter and light respectively, interact and exchange energy.

The mathematical formulation of QED is highly complex and involves the use of Feynman diagrams which allow physicists to calculate the probabilities of different particle interactions and make predictions about experimental outcomes.

QED has been incredibly successful in explaining and predicting a wide range of phenomena. It has been used to accurately calculate the magnetic moment of the electron, explain the behavior of atoms and molecules, and understand the interaction between light and matter in various experiments.

One of the most remarkable achievements of QED is its ability to explain the phenomenon of quantum electrodynamical vacuum fluctuations. According to QED, even in empty space, **virtual**

particles continuously pop in and out of existence, creating a constant background of quantum fluctuations. These fluctuations have been experimentally verified and have profound implications for our understanding of the fundamental nature of reality.

Furthermore, QED has provided a basis for the development of other quantum field theories, such as quantum chromodynamics (QCD) and electroweak theory. These theories extend the principles of QED to describe the behavior of other fundamental particles and forces, such as quarks and the weak nuclear force.

8.6 Quantum Chromodynamics

Quantum Chromodynamics (QCD) describes the strong interaction between quarks and gluons, which are the building blocks of protons, neutrons, and other hadrons. It is a crucial component of the Standard Model of particle physics.

The strong interaction, governed by QCD, is responsible for holding quarks together to form composite particles like protons and neutrons. Gluons, the mediators of the strong force, carry the color charge that characterizes this interaction. The inclusion of the word "color" in Quantum **Chromo**dynamics (QCD) might lead one to wonder if there is any connection between this scientific theory and the concept of colors.

While the term may seem peculiar, it indicates a different meaning within the realm of particle physics and the study of strong interactions.

In the context of QCD, the term "color" does not refer to the ordinary visual perception of colors. Instead, it is a metaphorical concept used to describe a fundamental property of quarks and gluons. I know, I know these physicists have a serious lack of imagination and an incredible skill in making people confuse.

While electric charge determines the behavior of particles under the electromagnetic force, color charge governs their interactions under the strong force.

Color charge in QCD is a unique property that comes in three distinct types or "colors": red, green, and blue. These colors are purely abstract labels assigned to the different types of quarks and gluons, similar to the way electric charge is labeled as positive or negative. The use of the term "color" aims to provide a convenient analogy to help visualize and understand the complex interactions occurring within the strong force.

A short example of the dynamics

In contrast to electric charge, which can be positive or negative, color charge is always conserved. This means that the combination of quarks and gluons in any given interaction must result in a neutral combination of colors.

Furthermore, in the field of quantum chromodynamics, quarks are known to exist in six distinct "flavors". These flavors are referred to as up, down, charm, strange, top, and bottom. What makes these flavors interesting is that each one carries a fractional electric charge. Furthermore, quarks interact with one another through the strong force, which is mediated by particles called gluons.

Gluons play a crucial role in the dynamics of quarks. They can be visualized as the "glue" that holds quarks together within atomic nuclei. The exchange of gluons between quarks enables the transfer of energy and momentum. This exchange mechanism is a fundamental process in the realm of particle interactions.

Through the exchange of gluons, quarks can effectively communicate with each other, altering their energy states and momentum. This phenomenon is akin to a conversation between particles, where information is exchanged through the intermediary of gluons. Just as we exchange ideas and thoughts through language, quarks engage in exchanges of energy and momentum via gluon interactions.

It is important to note that the fractional electric charge carried by each quark flavor determines the strength and nature of their interactions with gluons. This intricate interplay between quarks and gluons forms the foundation of QCD, offering insights into the behavior of matter at the subatomic level.

One of the remarkable features of QCD is its property known as asymptotic freedom. At high energies or very short distances, the strong force weakens, allowing quarks and gluons to move more freely. This behavior arises due to the peculiar nature of the strong force, where the interaction becomes weaker as the distance between particles decreases.

On the other hand, at low energies or large distances, the strong force becomes stronger, preventing quarks from existing as free particles. This phenomenon is known as confinement. It explains why quarks are always observed bound together within hadrons and cannot be isolated individually.

Quantum Chromodynamics has been extensively studied and verified through experiments, particularly at high-energy particle colliders such as the Large Hadron Collider (LHC). These experiments have provided crucial insights into the behavior of quarks and gluons and have helped confirm the predictions of QCD.

THE DARK UNIVERSE

To quote one of the famous and crazy theoretical physicists of our century Dr. Michio Kaku: *"Most of the universe is not made out of atoms. Every textbook is wrong, and every chemical class is wrong. Your high school chemistry teacher is wrong in saying that everything is made of atoms.*

We now know that most of the universe is made of Dark Matter and Dark Energy. Dark Matter makes up 25% of the universe and Dark Energy makes up 70% of the Universe. The stars make up 4%, and what about us? We make up 0.03 percent of the universe."

But what exactly is this physicist talking about? How is it possible that the universe we know is not made up of classical atoms? Centuries of study, thousands of equations and formulas, thousands of books and publications, for what? To discover that everything around us is made of something else entirely? What does this even mean?

In this chapter, we will give you an overview of the current theories and observations surrounding the so-called dark matter and dark energy, shedding light (better to say: shedding photons) on these mysterious phenomena that apparently shape the vast expanse of the universe.

9.1 The Mystery of Dark Matter

Despite its name, dark matter is not actually dark in color but rather refers to the fact that it does not interact with light. This enigmatic substance cannot be detected directly, making it even more intriguing.

Its existence was first proposed based on the gravitational effects it has on visible matter, such as stars and galaxies.

In fact, dark matter is estimated to constitute a large percentage of all matter in the universe, while ordinary matter, including the atoms that make up everything we see, accounts for only a small fraction of what our eyes perceive. So, what exactly is dark matter? The truth is, we don't know for sure yet.

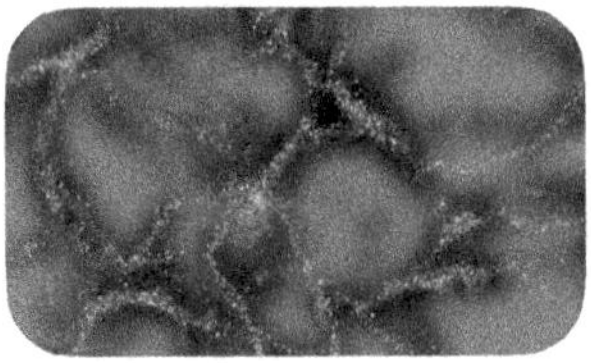

Dark matter, whatever it is, must be plentiful, cold, and free of collisions. Dark matter's composition is of interest to astronomers because it sheds light on galactic dynamics and cosmic development.

This is mostly an astronomical issue, but as physics applies to everything, this common and mysterious material may even have illuminating quantum characteristics.

The cosmic microwave background, the bending of light around massive structures, the rotation curves of galaxies, the temperatures of galaxy clusters, the large-scale structure of the universe, and countless other pieces of evidence all point to the existence of an undetectable matter. Therefore, its existence is without dispute.

We have no idea what it is, but we have a good idea of what it isn't!

We know this isn't just regular, classical stuff like rocks and planets since gravitational effects can't be explained by the amount of normal matter in the cosmos as measured in the early universe.

Also, as previously said, black matter must be both cold and collisionless. Since dark matter is both cold and collisionless, meaning it does not interact with either itself or ordinary matter, we may infer that it is pervasive throughout the universe.
You are, in fact, surrounded by a sea of dark matter particles, and they're flowing through your room at this very moment.
However, we can only detect dark matter indirectly, as it doesn't interact with ordinary matter.
An example would be the so-called "rotation curves," which show the relationship between the luminosity of a galaxy and the speed with which its stars orbit it.
According to this first principle of Newtonian physics, the density of a substance decreases as one moves away from its center and the majority of its mass is concentrated in its center. Furthermore, the furthest stars shouldn't be circling at a high rate of speed due to their great distance from the center.

Still, that has nothing to do with what we saw! Stars in galaxies are really moving about the center of their respective universes at a far quicker rate than what would be expected if all the luminous stuff were to be considered. This contradicts our previous observations. Stars' motion and velocities are affected by other forces, such as an unseen form of matter and, consequently, gravity.

All this mystery only increases the desire among theorists to understand as much as possible of this phenomenon, making up their own pet theory of what dark matter is and how it might work. We have candidates like:

- Weakly interacting massive particles (WIMPS);

- Self-interacting dark particles;

- Axions or Axion-like particles;

- Primordial Black holes;

- Sterile neutrinos.

And the list goes on and on. However, the most important thing is that all of these diverse theories are able to anticipate how dark matter would behave in our universe by their predictions. An evaluation is currently being conducted to determine which theory is the most accurate and closest to the truth.

It has been hypothesized various and different possibilities of how to detect dark matter, for example: it has been suggested that two particles of dark matter might occasionally collide and destroy each other in a burst of gamma-ray radiation that is visible to us.

Or it could be that dark matter buries itself deep in the heart of a star and raises the temperature higher than what you might normally expect. We might even detect it here on the Earth, we could set up a cryogenic ultra-cold detector and wait a really long time, and then occasionally a dark matter particle will hit the detector and heat it up just a little bit and detect that hit.

To visualize them better we have developed dozens of detectors and instruments and observatories around the world hunting for dark matter, every single second of every single day. But we just haven't seen any yet.

Scientists around the world are engaged in an intense pursuit to understand the enigmatic nature of dark matter. With its pervasive presence in the universe, dark matter remains an elusive puzzle that has captivated the minds of physicists and astrophysicists alike. To shed light on this mysterious substance, researchers have embarked on a relentless quest, utilizing state-of-the-art methodologies and revolutionary technologies.

At the forefront of this scientific endeavor are the proponents, as mentioned, of the WIMP theory, which posits that dark matter particles interact solely through gravity. These particles, aptly named Weakly Interacting Massive Particles, present an enormous challenge when it comes to detection. Unlike ordinary matter, which interacts through electromagnetic forces, WIMPs barely leave a trace as they traverse the cosmos.

The quest to uncover the true nature of dark matter has led scientists to devise ingenious methods and employ cutting-edge technologies. Advanced detectors, such as underground experiments and space-based observatories, have been meticulously designed to capture even the faintest signals that could hint at the presence of dark matter. These detectors are painstakingly shielded from cosmic rays and other sources of interference, allowing researchers to sift through an overwhelming sea of data in search of elusive clues.

While the search for dark matter may seem like an insurmountable challenge, the dedication and perseverance of scientists continue to drive progress. Collaborative efforts, international collaborations, and interdisciplinary research are bringing together the brightest minds from various fields to tackle this cosmic conundrum.

9.2. Dark Energy

Let's continue our tour with one of the most mysterious and thought-provoking phenomena in the universe. Dark energy. It's believed to be the force behind the accelerated expansion of the universe. It's called "dark" because just like dark matter, it does not interact with light or any other form of electromagnetic radiation.

Despite its elusive nature, scientists have been able to study its effects on the cosmos through careful observations and measurements. These studies have led to the realization that dark energy constitutes about 69% of the total energy content of the universe. Isn't that baffling? This discovery challenges our understanding of the fundamental forces and particles that govern the universe.

How does Spacetime Expand?

It was in 1929 when Edwin Hubble conducted an investigation into the phenomenon of the wavelength of light that is released by distant galaxies shifting towards the red end of the electromagnetic spectrum as it travels through space. He discovered that galaxies that were fainter and farther away had a greater degree of redshift, but galaxies that were closer to us did not exhibit as much of a shift.

Hubble eventually came to the conclusion that this was due to the fact that the cosmos itself was expanding, which caused the redshift to take place. This is because the wavelength of light gets stretched as the universe expands. Consequently, the frequency decreases, and the color shifts. It was formerly believed that the force of gravity would cause the expansion to either slow down or even retract. However, this was not the case.

The expansion of spacetime can be visualized using the analogy of an inflating balloon. As the balloon is inflated, the distances between any two points on its surface increase. According to the theory of general relativity, the fabric of spacetime is not rigid but rather flexible, allowing it to stretch and expand. This expansion occurs uniformly in all directions, causing the universe to expand isotropically.

It's important to note that the expansion of spacetime does not imply that objects within the universe are physically moving away from one another. Instead, it is the fabric of spacetime itself that stretches, carrying the galaxies along with it. As the balloon expands, the dots move away from each other, even though they are stationary relative to the balloon's surface.

Dark energy is thought to exert a repulsive force on matter and energy, causing the universe to expand at an accelerating rate.

This expansion, in turn, influences the propagation of light and other forms of energy through the spacetime grid. As dark energy continues to push the boundaries of the universe outward, it simultaneously shapes and stretches the fabric of spacetime.

Imagine the spacetime grid as a vast network of interconnected nodes and threads. Dark energy acts as a force that constantly adds tension to these threads, causing them to stretch and create new connections. This continuous stretching and creation of connections result in the expansion and evolution of the spacetime grid.

What is this energy made of?

The qualities of space do not change as it expands; rather, there is just more of it around. Because galaxies are tightly bonded groups of things that are kept together by gravity, we do not feel this dilatation in our day-to-day lives. However, new space is continually being generated everywhere.
Every second, new things are being created in every part of the cosmos where there is some empty space. Consequently, it would appear that dark energy is a form of energy that is inherent to empty space. It is energy that is more powerful than everything else that we are aware of, and it is steadily becoming more powerful as time goes on.

In other words, the energy of empty space is greater than the sum of all the other energy in the cosmos. There are a number of different hypotheses on the nature of dark energy.
There is a theory that suggests that dark energy is not a physical entity but rather a characteristic of space. Empty space is not nothing; rather, it possesses its own life force.

So much so that space may be generated by it, and it is highly active. As the cosmos continues to expand, it is possible that the gaps are simply being filled by an increasing amount of space, which ultimately results in the universe expanding at a quicker rate. The concept that empty space is actually filled with transitory, virtual particles that spontaneously and continuously develop from nothing and then dissolve into nothing again is another theory that has been proposed.

There is a possibility that the energy of the particles is dark energy. It is also possible that the generation and destruction of these particles are the result of the undulatory motion of a quantum field of some type.

It is also possible that dark energy is a type of dynamic energy fluid or field that is unknown to science. This type of energy penetrates the whole universe, yet it has an influence on the cosmos that is diametrically opposed to that of conventional energy and matter.

In the event that it does exist, however, we do not know how, where, or how we may discover it.

We don't know also how these forces would determine the ultimate fate of the universe. Is the universe going to expand forever without ever stopping? Expand and stop at some unfathomable distant time in the future?

Or expand for a while, then the expansion will be overcome by gravity and have the universe collapse in some sort of big crunch? Unfortunately, nobody knows the answer and the debate rages.

While the exact mechanisms through which dark energy interacts with the spacetime grid are not fully understood, the concept provides a compelling explanation for the continuous expansion of the universe. The interplay between dark energy and the spacetime grid is a topic of ongoing research and exploration, as scientists strive to unravel this mystery.

9.3 Antimatter

There is no gold, no oil, and not even diamond that can be considered the most valuable object in our universe. It must be antimatter. Whatever the exorbitant expense of this item is well worth it because it has the potential to provide a solution to the riddle of why anything that exists in our universe exists at all.

In our planet and in the universe in general, every particle has a counterpart known as an anti-particle. These anti-particles are similar to the particles in every aspect, but they have the opposite charge and spin.

For instance, an electron has a **positron** as a counterpart (which is an electron with a positive charge) and a proton has an anti-proton, and so on. Crazy right?

When a particle comes into contact with its anti-particle partner, that is to say, when the matter comes into contact with anti-matter, the two particles have the potential to pair-annihilate, meaning that they fully cancel each other out, leaving just two photons to take away the energy. And this also works in the opposite direction!

The creation of particle and anti-particle pairs is possible with the use of pure radiation. In point of fact, we imagine that this is how the very first particles were generated at the very beginning of the universe.

If matter and anti-matter are constantly formed in pairs (as physicists claim), then there should have been precisely the same number of each at the beginning of time. However, this is not the case, there's so much matter around us - and we have to wonder to what extent the anti-matter exists.

A good question would be: Why does exist any matter at all? Shouldn't everything have just been demolished once more, leaving behind nothing but a light that was bathed in a vacuum?

It would appear that the solution that is most likely to be correct is that the cosmos began with a somewhat greater amount of matter in comparison to anti-matter. If there were a little greater number of matter particles than anti-matter particles, then just much of everything would have been destroyed.

This would have resulted in a world that was dominated by photons, with only a very small number of particles that were unable to find a partner for annihilation.

As a result of the fact that there are about one billion times more photons than there are particles of matter, we have estimated that just one particle of matter survived for every billion particles of matter that were destroyed.

The question that has to be answered is why particles were formed with such a one-in-a-billion excess in comparison to anti-particles. It would appear that the method in which the universe interacts with particles is fundamentally distinct from the way in which equivalent particles interact with the universe.

A significant number of physicists believe that the explanation may be found in the fundamental symmetries of the cosmos, or more specifically, in the brokenness of these symmetries.
There's no definitive answer to this question to this day; nevertheless, one definite thing is definite is that the cosmos conceals variables and most likely some more basic principles that lie beneath the surface and are utterly beyond our comprehension at this point in time. If we are to have any hope of having a thorough knowledge of this system, it will take us decades, if not centuries, of research and observations.
We have no other option than to produce some anti-matter, examine it, and put it through its paces in order to get a deeper comprehension which is not precisely a simple procedure to carry out. The observation of anti-particles is rather simple.

For instance, positrons (also known as anti-electrons) are produced in nature on a regular basis, their can occur in the Sun, in any radioactive decay, or when cosmic rays strike the atmosphere. This observation led to the discovery of anti-matter in the first place. Furthermore, particle accelerators have the capability of producing more "*exotic*" antimatter, such as anti-protons, just by combining conventional matter with one another.

9.4 Creating Antimatter

Particle Accelerators

The creation of antimatter particularly in particle accelerators is an intriguing and complex process that involves the manipulation of particles and their interactions at incredibly high energies.

Particle accelerators, play a vital role in the production of antimatter. These colossal machines are designed to accelerate particles, such as protons or electrons, to extremely high speeds, close to the speed of light. By doing so, the particles gain an enormous amount of kinetic energy.

The first step in creating antimatter involves generating a beam of particles, typically protons, within the particle accelerator. These protons are then accelerated to reach energies in the order of teraelectronvolts (TeV) or even higher. Once the particles have gained sufficient energy, they are directed towards a target material.

When the accelerated protons collide with the target material, the protons interact with the protons or nuclei of the target material, releasing an immense amount of energy. This energy can be converted into new particles, including antiparticles.

The collision between the accelerated protons and the target material leads to the creation of both matter and antimatter particles. For example, when a proton collides with a proton, they can annihilate each other, resulting in the production of various particles, including antiprotons. Similarly, when a proton collides with a nucleus, it can give rise to antiprotons as well as other antiparticles like antineutrons or even antihydrogen atoms. Once the antimatter particles are created, they need to be carefully captured and stored.

Storage and Containment

It is difficult to maintain antimatter around for an extended period of time since it is extremely reactive and can be destroyed when it comes into touch with conventional matter. This is because antimatter is surrounded by matter. In order to find a solution to this problem, scientists have explored novel approaches to the storage and containment of antimatter products.

The use of electromagnetic fields to capture and suspend antimatter particles is one method that may be utilized. This method prevents the particles from coming into touch with the matter that is in the surrounding area. Another technique involves the utilization of magnetic bottles, which employ magnetic fields to trap antimatter and so prevent its annihilation.

Instead, when it comes to whole anti-matter atoms, which are electrically neutral and hence impossible to even store using electric and magnetic fields, the task becomes very challenging.

An expensive and complex method employed for the containment of anti-hydrogen for example is the use of cryogenic technologies. Anti-hydrogen is typically stored at extremely low temperatures, to reduce the thermal energy of the particles and minimize their interactions with the surrounding matter. Cryogenic systems, such as liquid helium or liquid nitrogen baths, provide the necessary conditions for preserving anti-hydrogen and preventing premature annihilation. Furthermore, scientists have also explored the possibility of utilizing sophisticated vacuum chambers for the storage of anti-hydrogen. By creating a vacuum environment, the presence of matter particles can be significantly reduced, minimizing the chances of annihilation. These vacuum chambers are carefully designed and maintained to ensure optimal conditions for anti-hydrogen storage.

It is important to note that the storage and containment of anti-hydrogen require highly controlled environments and rigorous safety measures. The potential energy release during an accidental annihilation event could be catastrophic. Scientists continuously work on improving storage techniques and implementing advanced safety protocols to minimize any risks associated with handling anti-hydrogen.

To give you an idea of the amount of work that is required, for the most basic sort of anti-matter atom, which is precisely anti-hydrogen - composed of a single anti-proton and a positron, as opposed to the proton and electron that are found in a standard hydrogen atom - the National Aeronautics and Space Administration (NASA) calculated in 1999 that the cost of a single gram of anti-hydrogen was 62.5 trillion dollars when all expenditures were taken into consideration.

Unfortunately, we are going to wait a little bit before we can start using antimatter engines to power our starships.

And you who thought that oil was expensive.

THE QUEST FOR A THEORY OF EVERYTHING

The Quest for a Theory of Everything is a captivating journey undertaken by scientists and physicists to uncover a unified theory that explains the fundamental forces and particles of the universe. This pursuit aims to bring together the principles of quantum mechanics, which govern the microscopic world, and general relativity, which describes the behavior of gravity at a macroscopic level.

The quest begins with the realization that these two pillars of modern physics are incompatible, each successfully explaining certain phenomena but failing to account for others. When attempting to apply both theories simultaneously, inconsistencies arise, highlighting the need for a more comprehensive framework.

The reconciliation of the macroscopic world, which is driven by gravity, and the microscopic world, which is dominated by quantum effects, is one of the most significant issues that arises in the search for a Theory of Everything.

Particles may exist simultaneously in numerous states, showing qualities such as superposition and entanglement, which are both characteristics of the quantum world that work on a scale that allows that to happen.

On the other hand, general relativity describes objects positioned in precise coordinates and treats time as a "geometric" dimension and gravity as the curvature of spacetime caused by massive objects. Bridging these two seemingly contradictory descriptions is a formidable task.

This is a contradictory and difficult task because it would imply that time as well as gravity at the microscopic level would have to be quantized, a thing that remains to be proven.

Numerous theoretical frameworks have been proposed in the quest for a Theory of Everything such as the String theory, in which the fundamental building blocks of the universe are not point particles but rather tiny, vibrating strings. Or Loop quantum gravity, which seeks to quantize spacetime itself. In this framework, spacetime is viewed as a network of interconnected loops, representing discrete units of space and time. It is important to note that while these theories show promise, they are currently highly speculative, but let's examine them better.

10.1 String Theory and Quantum Gravity

String theory is a comprehensive explanation of the fundamental nature of the universe. It's one of the theoretical frameworks that aims to reconcile the principles of quantum mechanics and general relativity. Similar to the Quantum Field Theory, String Theory is a way to model the universe in a useful mathematical structure.

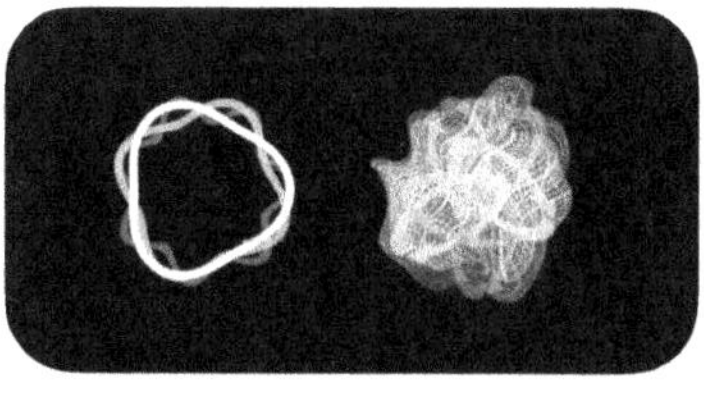

String theory is a model that proposes that the tiniest building components of the cosmos are not zero-dimensional point-like particles, but rather minuscule, one-dimensional "strings" that vibrate at a variety of frequencies and in a variety of dimensions.

A comprehensive description of the many kinds of matter, forces, and energy that exist in the cosmos may be derived from the various combinations of frequency and dimensions.

The Quantum Field Theory, on the other hand, postulates (as we have seen in Chapter 8) that all particles are excitations in fields that cover the whole universe, and that forces are the result of the interaction of these particles, which is mediated by other particles known as Bosons. On the basis of this framework, our most comprehensive and precise understanding of the universe is collected in what is commonly referred to as the standard model of particle physics.

Both String Theory and QFT are both based on the principle of quantum mechanics and special relativity. QFT is valued for being the most accurate theory we ever have - its predictions concur with experimental results in some cases to 10 parts in a billion.

So why do we even bother with the crazy idea of a string *thing*? The biggest reason is that it so happens that a certain combination of a closed string vibrating in 10 dimensions can model a particle that looks identical to what is called a **Graviton**.

The Graviton is a theorized missing particle of the standard model that would carry the gravitational force, similar to the way photons carry the electromagnetic force, W and Z Bosons carry the Weak Force and the Gluons the Strong Force.

Keep in mind that Gravitons are theoretical only, they've never been detected, and they're predicted only by string theory. The conceptualization of this new particle was the reason that made this theory so popular some decades ago.

Physicists thought that it could provide the mathematical leap we needed to integrate gravity into Quantum Field Theory. And if that could be done, it would solve the Holy Grail of physics: the quantization of gravity.

The issue that arises when attempting to quantize gravity within the framework of quantum field theory is referred to as *Renormalization, which* is the procedure of making a theory valid to an infinitesimally small length.

We obtain infinities in the equations when we try to achieve this quantization (which is not what we want, quantum physics is all based on discrete portions).
So quantizing gravity using QFT, doesn't really work.
It's possible to make an attempt to explain this concept with the following analogy: spacetime is the backdrop on which quantum fields exist, just as paint occurs on a canvas. As a result, the painting falls apart when an attempt is made to transform the canvas itself into a painting. All crumbles when we try to quantize the canvas.

In contrast to QFT, String Theory provides a solution to the problem of quantizing gravity. However, there is a significant cost that we must bear in order to be able to do so, and that is the fact that we must deal with 6-7 additional dimensions with a total of 10 or 11 dimensions. The theory itself implies that these theories have forms and manifest themselves in a multitude of dimensions.
This is a challenge due to the fact that these dimensions have never been located. In addition, this is not due to a lack of effort on the part of the scientists; they have tried and tried again to identify the existence of these other dimensions, but they have always been unsuccessful.

At this point, you could be thinking, "Well, then we ought to get rid of it and come up with some other kind of crazy idea." Nevertheless, the fact that String theory is not able to discover these additional dimensions does not necessarily mean that it is flawed.

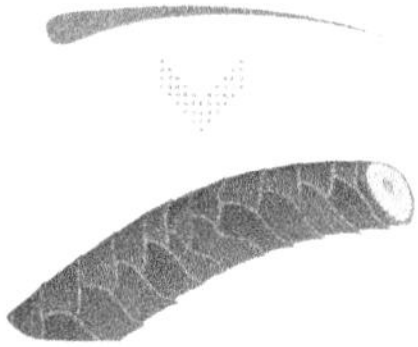

Explanations were developed by scientists on the possible locations of the dimensions that were missing.

One is *Compactification,* which can be thought of like this: imaging holding a little hair - from far away it looks like a thin almost one-dimensional line. But as you look at it more closely, you realize that is not just a line, it has an intricate three-dimensional structure with its scales, its Cortex, etc. The idea is the same with String theory, we see objects from a point of view where things look three-dimensional, but if you could zoom in, incredibly close, closer than we are technically capable of doing today, we would discover compact dimensions that are hidden from us.

The basic premise is that there could be a fourth spatial dimension separate from the other dimensions in the cosmos and that there might be more dimensions to the world than the ones we can perceive.

If we wanted to observe these other dimensions, we'd have to go outside our local membrane. The issue is that we are unable to test it with our present technology due to the enormous amounts of energy required to detect dimensions on such a microscopic scale. But this in no way disproves the validity of higher-dimensional spaces or mathematics.

In the past for example, gravitational waves were predicted by Einstein in 1916 but not detected until 2015, it took a full century to actually have the evidence. Perhaps we can navigate the complexities of string theory in the same fashion.

The world would be shocked if we could find these other realms! If this is true, it would provide strong proof that String theory, not QFT, is the best model of the cosmos. Such an approach would be analogous to how we discovered that General Relativity provides a more accurate description of gravity than Newton's equations, which are only approximate.

But hey! Could the two theories, QFT and String theory be combined to make a kind of composite theory? It doesn't seem so. The main problem is the extra dimensions. But are there any similarities? In QFT we define particles as point-like fuzzy vibrations having negligible dimensions. They have no defined size, they're excitations in the field that exist in four dimensions (3 spatial and 1 temporal).

In String theory, a particle is a one-dimensional object vibrating in at least 10 dimensions. It has a length, but not width or height, it can be an open or closed string, similar to a rubber band.

This 1D string can have a size, it is near the Planck scale, which is about 10^{-33}cm. We can think of this as analogous to the strings in a guitar. It can vibrate to make sounds and just like the frequency of the vibration determine the pitch of the sound, the frequency of strings in String theory determines the type of particle it is. For example, a low-energy vibrational mode of a string may correspond to an electron, while a higher-energy mode may represent a photon.

Thus, the way and size of a string's vibration might cause it to behave as various particles. The amplitude of the vibrations in string theory is proportional to the number of particles, just as it is in QFT. This is about as far as the similarities go.

It is unusual to many physicists that one of the most essential forces in the world is an exception to the rule of quantum mechanics, which is why they are certain that gravity must also be quantum.

The behavior of gravity at extremely small distances or high energies, such as those encountered near the center of a black hole or during the early stages of the universe, is currently unknown.

To understand these extreme conditions, a quantum theory of gravity is believed to be necessary. Such a theory would provide insights into the behavior of gravity in the most extreme environments and potentially answer fundamental questions about the nature of spacetime itself.

Currently, quantum field theories have been able to adequately explain three of the four basic forces: electromagnetism, the strong nuclear force, and the weak nuclear force.
But there is still no consistent quantum mechanical explanation of any force other than gravity. In their pursuit of a theory of everything, physicists are working toward a unification of all the basic forces through the development of a quantum theory of gravity.

You might be asking, why does this theory need all these dimensions? The simple answer is that its mathematics does not work in our 4 dimensions. For your sake, I'll spare you this math, but this string scenario only works if we allow the strings to vibrate in at least 6 extra dimensions or at least 10 total dimensions.

Introducing multiple dimensions could help in:

1. Resolving the Hierarchy Problem

The hierarchy problem refers to the stark difference in the strengths of the fundamental forces. Gravity, for instance, is significantly weaker compared to the electromagnetic and strong nuclear forces. String theory suggests that the extra dimensions could be responsible for this hierarchy. The additional dimensions may be "compactified" or curled up in such a way that they are too small to be directly detected. However, their presence influences the behavior of the fundamental particles, resulting in the observed force strengths.

2. Explaining the Origin of Mass and Particle Properties

Finding out where mass comes from is one of physics' great puzzles. The Higgs mechanism is presented in the Standard Model of Particle Physics to describe the process by which particles gain mass. Particle masses and other features in string theory instead, may be generated in large part by the extra dimensions. The masses of the particles and their unique characteristics, like gravity, are determined by the way the strings vibrate in these extra dimensions.

Before we go any further, it's worth noting that more dimensions are theoretically possible in quantum field theory, but they haven't been necessary for the theory to function thus far; the math holds up well in the four dimensions we can see, provided that we disregard gravity.

I wouldn't hold it against you if you believe this whole multi-dimensional confabulation is insane!

Here we're trying to solve a really hard problem, probably the hardest in all of physics. At the moment we can't afford to throw out any odd or eccentric-sounding theory just based on intuition or incredulity. We should evaluate all crazy ideas based on their merit.

Unfortunately, after all this description, String theory doesn't appear to be a great alternative because its predictions are either incorrect or untested. This is why it's getting out of favor by many physicists - on paper it nicely allows us to model gravity but in practice, it doesn't bring us any new answers. It's a good mathematical exercise but has limited physical significance.

10.2 The Enigma of Black Holes

Among the many mysterious and fascinating phenomena in the cosmos, black holes rank high. Nothing, not even light, can evade the enormous gravitational attraction of these celestial bodies. Picture a spot in space where the pull of gravity is so great that it forms a gravitational well that not even particles or electromagnetic waves can evade it. Where do they originate, though?

When a massive star reaches the end of its life, it undergoes a cataclysmic event known as Supernova. During this explosive phase, the star's core collapses under its own gravitational pull, leading to an incredibly dense and compact object called Neutron Star. But let's proceed with order.

Stars are gigantic clumps of hydrogen atoms that formed when a gas cloud collapsed under the force of their own gravity. At its core, nuclear fusion transforms hydrogen into helium while simultaneously unleashing an enormous quantity of energy. The delicate equilibrium between the two forces is maintained by this energy's push against gravity in the form of radiation.

As long as fusion continues inside the star, it will stay stable. However, in stars much heavier than our Sun, the intense heat and pressure inside the core might fuse even heavier elements until iron is formed. The fusion process that forms iron does not produce energy, unlike all the elements that came before it.

The star's core collapses when the amount of iron accumulates up to a certain level, disrupting the delicate balance between radiation and gravity. A star collapses in the blink of an eye. Traveling at a speed close to 1/4 of the speed of light, adding additional mass to the core, and eventually exploding into a supernova.

The result is a neutron star or, in the case of very massive stars, a black hole formed when the core mass collapses.

If we look at a black hole, what we'd really be seeing is what is called the *Event Horizon* which is the boundary that divides the black hole from the rest of space; and it is what we distinguish when we stare at a black hole.

It is theoretically impossible for anything to escape the event horizon unless its speed is greater than light. So what we see is just a massive black sphere reflecting nothing.
But if the Event horizon is delimiting the black part, what is the "hole" part of the black hole?

The singularity.

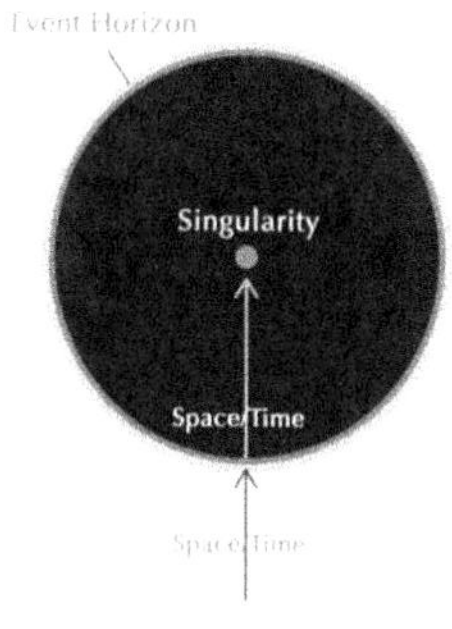

To this very day, we're not sure what singularity is. A singularity may be an incredibly small point volume with all the mass concentrated in it, or something completely different. The result though is an astonishingly strong gravitational force. This intense gravity distorts even the time around it.

Although black holes are famous for "consuming stuff", it's vital to remember that they don't really "suck" things into themselves in the same way a vacuum cleaner would. Instead, an accretion disk forms around them, which spirals into the black hole, as a result of their enormous gravity, which accelerates and absorbs matter and energy from the surrounding area. Because of the tremendous amounts of energy released as X-rays and other forms of high-energy radiation, astronomers are able to identify black holes through this mechanism.

There are black holes of varying sizes; some have stellar mass, meaning they're many times as massive as the Sun and as big as an asteroid and there are the *supermassive black holes* that have been devouring matter for eons and may be located near the galactic core. Google Supermassive black hole S5 0014+81 is currently the largest one known, with a mass forty billion times that of our Sun! Its diameter measures 236.7 billion kilometers.
But don't be afraid black holes, despite their immense strength, will eventually dissipate through Hawking radiation in due time.

As we saw previously, virtual particles - matter and anti-matter - pop into existence and annihilate each other again in empty space. This process occurs at the brink of a black hole, drawing one virtual particle into the hole while the other can escape and transform into a physical one. Energy in this way is being drained from the black hole.

This occurs quite slowly initially but speeds up as the black hole shrinks. It radiates at room temperature when it reaches the mass of a big asteroid, and roughly the sun's heat when it reaches the bulk of a mountain.
The black hole explodes in a massive explosion at the very end of its existence, radiating out energy equivalent to billions of nuclear bombs.
However, the pace of evaporation is so slow that the largest black holes known to exist may take an eternity to dissipate. This is so long that when the last black hole radiates away, nobody will be around to witness it.

Even though Albert Einstein postulated the existence of black holes in his general theory of relativity, the scientific world did not have a convincing proof for their existence until 1971, when a celestial object named Cygnus X-1 was discovered.

Black holes serve as laboratories for studying matter under extreme conditions. Near a black hole, matter experiences tremendous pressure, heat, and tidal forces due to the strong gravitational pull. Our comprehension of matter's basic characteristics can be enhanced by studying its behavior in these extremely harsh environments.
The function of black holes in explaining the Big Bang is also substantial. According to the Big Bang hypothesis, which is in line with the black hole theory, the cosmos started as a singularity. Insights about the Big Bang, the expansion of the universe, and the creation of cosmic architecture can be derived from research on black holes.

10.3 Loop Quantum Gravity

Loop Quantum Gravity (LQG) is another theoretical framework that aims to reconcile the physics of the tiny and the gigantic.

LQG provides a promising avenue for understanding the nature of space, time, and gravity at the microscopic level.

At its core, LQG suggests that space and time are not continuous and infinitely divisible, as assumed by classical physics, but rather discrete and quantized. This means that space and time are composed of tiny indivisible units or "quanta," similar to the way digital images are composed of pixels.

The term "loop" in Loop Quantum Gravity refers to the fundamental building blocks of space and time, which are known as "spin networks".

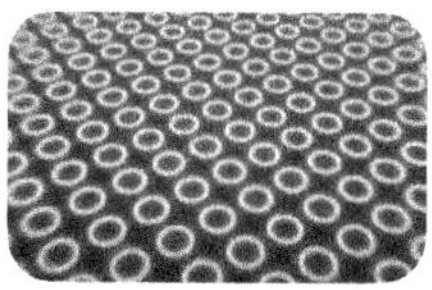

These spin networks are made up of interconnected loops, forming a complex web-like structure. Each loop represents an elementary unit of space, and the connections between loops represent the relationships between these units.

Quantum gravity, as suggested by LQG, treats gravity as a force that arises from the interaction of these elementary units of space. In LQG, gravity is not described by a continuous field, as in general relativity, but rather as discrete quantities associated with the spin networks. These discrete quantities, known as "quantum states," encode the properties of space, such as its geometry and curvature.

One of the most fascinating aspects of Loop Quantum Gravity is its introduction of the concept of "quantum jumps" in both space and time. Similar to how electrons can jump between different energy levels within an atom, LQG suggests that space and time themselves can undergo quantum jumps, transitioning between various configurations at the quantum level.

Although Loop Quantum Gravity is a complex and still developing theory, it holds great promise for addressing some of the fundamental questions in theoretical physics. It provides a potential framework for understanding the behavior of gravity at the quantum level, which is crucial for a unified theory of physics.

A dense and complicated web is one way to picture this network-like structure of interconnected loops; each loop represents a connection between nearby atoms in space. In the same way that a spider spins its web, these loops create a complex web of interdependence among the atoms in space.

The end product is general relativity recast in terms of extremely abstract building units, namely quantum circuits of the gravitational field rather than spatial-temporal chunks.

Success and Failure

The main achievement of loop quantum gravity is that it successfully merges the two widely accepted theories of general relativity and quantum mechanics while preserving their essential foundational assumptions. Also, without introducing major assumptions like the presence of strings, additional dimensions, or supersymmetry.

Although LQG exhibits background independence with respect to various 3D spatial geometries, whether or not this independence carries over to 4D spacetime remains debatable. Additionally, LQG does not address the issue of time.

In a broader sense, this theory can only be considered a success if it provides us with the classical limit equations of traditional general relativity applied to enormous, non-quantum scales. Even if it could, it's unclear how.

Research, however, is far from finished.

Both loop quantum gravity and the more well-known string theory remain trapped in their own theoretical quagmires, unable to establish any kind of experimental connection or interaction with the actual cosmos. Nevertheless, mathematical calculations have yielded fascinating hints about the universe's fabric, and that fabric, as you might have perceived, is quite odd.

QUANTUM COMPUTERS

How Does a Quantum Computer Work?

A quantum computer is a complex and fascinating machine that harnesses the principles of quantum mechanics to perform computations. Unlike classical computers that use bits to represent information as either a 0 or a 1, quantum computers use quantum bits or ***qubits***, which can represent both 0 and 1 simultaneously.

What is a Bit

If you want to grasp quantum computing, you must first be familiar with the term "*bit*" and have a general idea of how computers are built.

Now, for educational reasons let's focus just on one piece of these machines: the **chips.** Every single computer chip has modules, which have gates, which in turn have transistors.

A transistor is the simplest form of a data processor in computers, basically, a switch that can either block or open the way for information coming through.

Each bit, an abbreviation for "binary digit," may hold a value of zero or one. To a computer, it represents the most fundamental kind of information and serves as the fundamental building block of all digital data.

Generally speaking, if a current passes through the transistor this represents a "1" and if a current doesn't pass through this represents a "0". Digital logic circuits though, usually use voltage levels above some threshold to represent a "1" and below some (often lower) threshold to represent a "0", with anything in between being indeterminate. This is not a requirement, though.

Binary information is also transmitted using magnetic properties; the two different types of polarities are used to represent zeros and ones.

Because there are only two potential states for data representation and manipulation in computers, the idea of binary digits is well-suited for use with electrical circuits.

More complicated information is typically represented by grouping bits together in bigger units. One common way to represent a single numerical value or character of text is in terms of bytes, which are units of 8 bits. The 256 potential combinations of 8 bits enable for the representation of a vast range of characters and values.

A combination of transistors, on the other hand, compose a *logic gate* that performs also basic operations; for instance, an A.N.D. Gate type will produce an output of 1 if all of its inputs are 1, and an output of 0 otherwise. In turn, a combination of logic gates finally can form meaningful modules, say, for adding two numbers. And if you can add, you can multiply, and after that, you can do pretty much everything on a calculative level.

Do you remember superposition?

Contemporary most modern and expensive transistors often have a size of 15 nanometers, which is 500 times smaller than a red blood cell and approximately 8 times smaller than the diameter of the HIV virus!

Since transistors are getting down to the scale of individual atoms, electrons in the circuits may be able to quantum tunnel to bypass a blockage. So sometimes the very existence of a switch or a barrier doesn't prevent the flow of information.
Quantum computers employ qubits, which may be set to one of two values, as opposed to bits.
But in the quantum realm, the qubit can be in any combination of the two states at once, not just one. This is, as we already know, called superposition - however, it must choose between being vertically or horizontally polarized the moment we want to assess its value, for example by passing the photon through a filter.

Therefore, the qubit can't be predicted to be either 0 or 1 as long as it remains unseen since it's in a superposition of the two possibilities. But as soon as you start measuring it, it will settle into a discrete state. Superposition here is a real game changer.

How does superposition implement the calculating power in quantum computing?

When qubits are in a state of superposition, they can hold and manipulate a vast amount of information simultaneously, leading to exponential computational advantages over classical computers. This unique property is what makes quantum computing a promising field for solving complex problems that are beyond the capabilities of traditional computing methods.

By making use of the many potential states of qubits, superposition allows quantum computers to execute tasks in parallel. Computers transmit enormous quantities of data in the form of ones and zeroes, which are the building blocks of all the visual and auditory content you see on screens. This binary system of bits is the foundation of even the most advanced supercomputers on the

planet. Computing is made possible by the fast cycling of these little on/off switches between 1 and 0, allowing computers to encode music, pictures, and data at a rate of trillions of operations per second.
The things computers are capable of doing just by changing the states of bits are mind-boggling.

Imagine for a second that there is a vast library full of books, and somewhere in one of them, there's the formula for the youth elixir. Using a conventional computer approach you would need to manually check each book and register a yes or no until it was found. Using a quantum computer is like opening all the books at once and discovering the magic formula immediately.
The parallelism of the calculating process is what makes this computing implementation so important.

Since we are already in this realm, we cannot mention another phenomenon that occurs in quantum computers, which is the Entanglement. A connection that allows the two qubits to respond in real-time to changes in each other's states, regardless of their physical distance. This means that from the measurement of a single entangled qubit, the characteristics of its companions may be inferred immediately, eliminating the need to look.

So to recap a quantum computer:

- it is set up with a certain amount of qubits;
- it applies quantum gates to entangle them and manipulate probabilities;
- and then finally measures the outcome, collapsing superpositions to an actual sequence of 0s and 1s.

As you can well understand this whole process is supported by a delicate balance, that is, the isolation of the system from the external environment - and the difficulty in constructing a computer with these characteristics consists precisely in this.

To better grasp what I'm trying to say, let me briefly explain the concept of ***Quantum decoherence***.

A quantum system undergoes quantum decoherence when its interactions with its surroundings lead it to behave conventionally and lose its susceptible quantum features.
Interactions between quantum systems and their environments can cause quantum coherence to break down, which in turn can cause superposition to collapse and classical physics behavior to emerge.
To successfully develop and apply quantum technologies, it is crucial to regulate decoherence, a process with far-reaching ramifications for many branches of quantum physics.

As we learned in a quantum system, the wave function describes the probability distribution of its possible states. However, when the system interacts with its environment, the wave function collapses into a single state, resulting in the observation of a definite outcome.

A multitude of elements, including thermal fluctuations, electromagnetic radiation, and interactions with other particles or fields, contribute to the decoherence process. As a result of these interactions, the quantum system loses its coherence and becomes entangled with its surroundings.

Comprehending and delineating decoherence presents a formidable undertaking due to its utilization of complex mathematical formalizations and the necessity to account for the intricate dynamics of the quantum system and its surroundings.

Nevertheless, significant advancements have been achieved by scientists in the investigation and regulation of decoherence, which has facilitated the creation of quantum error correction codes and additional methodologies aimed at mitigating its effects.

Quantum Computers

Quantum computers are poised to revolutionize the way we live and interact with technology. While we are still in the early stages of development, the potential applications of quantum computing are vast and exciting.

One area where quantum computers can significantly impact our daily lives is in the field of cryptography.

Cryptography, which is the practice of secure communication, relies on complex mathematical algorithms to encrypt and decrypt sensitive information. Nevertheless, the advent of quantum computers presents a substantial peril to the security of conventional encryption techniques.

Public-key cryptography, which also goes by the name asymmetric encryption, is a cornerstone of contemporary cryptography.
To ensure the security of encrypted data, this encryption method employs the computational difficulty of specific mathematical problems, such as factoring enormous numbers. Quantum computers, on the other hand, are capable of resolving these mathematical issues considerably more rapidly than classical computers.
The vulnerability resides in the potential for a quantum computer of adequate power to breach the prevalent encryption algorithms, thereby jeopardizing the security of sensitive data such as financial transactions, government communications, and personal information.

This has raised concerns among experts in the field, as the rapid advancement of quantum computing technology brings the reality of such attacks closer.

To address this threat, researchers and cryptographers are actively exploring new cryptographic algorithms that are resistant to attacks from quantum computers. These post-quantum cryptography methods aim to provide secure encryption even in the face of quantum computing capabilities. The development and adoption of quantum-resistant algorithms are crucial to ensure the long-term security of sensitive information.

Virtual signatures, which serve to validate the integrity and authenticity of digital documents and transactions, may also be susceptible to exploitation by quantum computers.

The risk of forged digital signatures increases as the power of quantum computers rises, which may result in identity theft, fraudulent activities, and other malevolent acts.
Ongoing endeavors are focused on the development of digital signature schemes that are resistant to quantum computer attacks. The objective of these schemes is to generate tamper-resistant and secure digital signatures that retain their validity despite the presence of formidable quantum adversaries.

But I want to reassure you with:

the first major algorithm that was developed for quantum computers is Shor's algorithm, named after Peter Shor a brilliant mathematician from the MIT. He developed an algorithm capable of efficiently determining prime factors for extremely large numbers.
But why should this concern us? Because that forms the foundation for 'RSA encryption', which is essentially the method used for all online transactions.

The decision to implement this encryption scheme was driven by its near-impenetrability to classical brute force algorithms. Cracking it with regular computers would require an astronomical amount of time and consequently energy, but a quantum computer could accomplish the task much more efficiently, potentially in a matter of days or even hours.

However, the construction of such a quantum computer is still pending. To perform such a complex calculation, a quantum computer would need over a million qubits.
So we're not going to be using quantum computers to break all internet security right now (if that's your intention).
It's highly unlikely that we'll witness any quantum cyber wars in the coming months or years.

Currently, the highest number of qubits achieved is 400+ by IBM, which is a significant difference compared to the range of 400 to a million.
However, certain researchers at IBM who specialize in quantum physics firmly believe that our world is undergoing rapid transformation. They predict that in the decades to come, people will transition to quantum-safe algorithms.
The potential risks associated with this shift are significant, but so are the opportunities for developing intelligent and advanced defenses.

Quantum computing applications

Quantum computers possess immense potential to revolutionize various domains of scientific research. With their unique computing capabilities, they can greatly impact a vast range of sectors. Here are a few examples:

Drug Discovery

Quantum computers have revolutionized the field of drug discovery by offering scientists a powerful tool to simulate molecular interactions and optimize drug designs. This cutting-edge technology has tremendous potential to accelerate the development of new medications and treatments.

Traditional computers struggle to accurately model the behavior of atoms and molecules due to their complex nature and the vast number of possible interactions. However, quantum computers can handle these challenges effortlessly, offering researchers a more comprehensive understanding of molecular interactions.

With the ability to simulate molecular interactions, scientists can explore countless drug design possibilities in a virtual environment. They can analyze the interactions between potential drug candidates and target molecules, predicting their efficacy and potential side effects. This allows researchers to focus their efforts on the most promising candidates, saving valuable time and resources.

Through iterative simulations, scientists can fine-tune the molecular structures of potential drugs, enhancing their potency and reducing potential toxicity. This iterative process can be time-consuming and computationally demanding for traditional computers, but quantum computers excel at these optimization tasks, helping researchers develop safer and more effective medications.

The accelerated drug discovery process facilitated by quantum computers has the potential to revolutionize healthcare. It can significantly reduce the time and cost required to bring new medications to market, making life-saving treatments more accessible to those in need. Moreover, by expediting the development of drugs, quantum computers can also aid in addressing emerging health crises and combating drug-resistant diseases.

Machine Learning and Artificial Intelligence

Artificial intelligence (AI) and machine learning are two areas where quantum computing shows a lot of potential. Quantum machine learning algorithms possess the capability to efficiently process and analyze large volumes of data, surpassing classical algorithms. This capability has the potential to drive progress in a wide range of fields, such as data analytics, pattern recognition, and AI model optimization. By harnessing the power of quantum computers, AI models can be trained more effectively. This is achieved through the ability to generate and explore numerous possibilities simultaneously, resulting in rapid and precise learning.

By leveraging the properties of quantum systems, such as superposition and entanglement, quantum computers can explore multiple possibilities simultaneously. This enables them to evaluate and analyze different scenarios in parallel, leading to faster and more accurate learning. This quantum advantage in exploring multiple possibilities can greatly enhance the training of AI models, allowing them to learn from a wider range of potential solutions and make more informed decisions.

Optimization plays a crucial role in machine learning, as it involves identifying the most optimal parameters and configurations for a given problem. Traditional optimization methods can face difficulties in locating the overall best solution because of the existence of numerous local optima. Nevertheless, quantum algorithms, like quantum annealing and variational algorithms, present different approaches for optimization that have the potential to overcome these challenges and discover more optimal solutions.

So, the integration of quantum computing into machine learning and AI holds immense promise. It enables us to process and analyze vast amounts of data more efficiently.

Finance and Portfolio Optimization

As you can well imagine, the financial industry can greatly benefit from the power of quantum computing. Quantum algorithms can tackle complex financial problems, such as portfolio optimization and risk analysis, by considering numerous factors and variables simultaneously.

This can lead to more efficient investment strategies, improved risk management, and better asset allocation. Furthermore, quantum computing can contribute to the development of advanced financial models and simulations, enabling financial institutions to make better-informed decisions.

Traditional methods for optimizing investment portfolios often rely on simplifications and assumptions that may not accurately reflect the complex nature of financial markets. Quantum algorithms, on the other hand, can consider a vast number of variables, such as asset performance, market trends, and risk factors, all at once. This capability can lead to the development of more robust and diversified investment strategies, maximizing returns while minimizing risks.

Financial institutions are constantly faced with the challenge of quantifying and managing risks associated with their investments. Traditional risk models often struggle to capture the complexities and interdependencies of various risk factors. With quantum computing, it becomes possible to analyze a multitude of risk factors simultaneously, including market volatility, credit risk, and operational risk.

With the development of advanced financial models and simulations, financial institutions can make better-informed decisions, ultimately driving innovation and growth in the industry.

Ultimately, quantum computing holds immense potential to completely transform a wide range of fields and industries and with ongoing advancements in research and development, the field of quantum computing holds the promise of pushing boundaries and finding solutions to currently unsolvable problems.

QUANTUM CONSCIOUSNESS

Although the field of quantum physics is enormous and ever-expanding, we hope that this book has equipped you with all of the concepts you need to comprehend and enjoy the quirks and follies of the microscopic particle world. We can now say that we are approaching the end of this intense journey for beginners, but before concluding, I would like to bring to your attention an interesting interpretation of quantum mechanics, which although with its controversies, is worth being presented and analyzed. We're talking about:

Consciousness in Context

Do you remember what we said at the end of Chapter 2? We said that the collapse of the wave function and so of the superposition is not something that has to do with a wacky conscious force emanating from the mind of the observer but rather a matter of measurement (more precisely, whether it's present or not).

So we cannot manifest reality if we focus really hard using our quantum mechanics power? Absolutely not.

BUT.

Why did some of the pioneers of quantum theories believe that consciousness and quantum mechanics were closely connected?
Even though physicists have made remarkable progress in understanding the intricacies of the microscopic world and gained

widespread recognition in the scientific community, it is fascinating to see that even after almost a century since the establishment of quantum mechanics, there is still an ongoing debate among physicists regarding the interpretation of its equations.

Nevertheless, it is undeniable that the enigmatic nature of the quantum realm has sparked unconventional theories that have ventured into the realm of what we might call mysticism.
One particularly prevalent idea is that consciousness has the potential to directly impact quantum systems and, consequently, influence our everyday reality. But, let's take a moment to explore the origins of this idea and how certain individuals believe that quantum theory provides evidence for it.

Understanding how the wavefunction collapses is crucial because our classical world consists of objects with well-defined properties, rather than probability clouds or functions. So, at what point does the transition from the quantum realm to the classical world occur?

Let's examine the process of the double-slit experiment once again, we can observe that the electron wave function travels through both slits, eventually reaching the electronic detector. And there, it stimulates the excitation of a second electron on the electron screen. That second electron initiates a cascade - an electrical impulse that travels along circuits to be detected by a computer, which then updates an image on a computer screen to display the location of the electron's impact.
That data then is carried by photons to light-sensitive molecules in our retinas, which in turn trigger electrical impulses in our visual cortex. Later on, further electrical signals in other brain regions provide an individual's subjective perception of the original electron's selected location on the screen.

This sequence of information from the detector to the mind is known as the *Von Neumann chain* (inspired by the renowned

Von Neumann

Hungarian-American physicist John Von Neumann). **He suggested that the wavefunction collapse occurs somewhere between the measuring apparatus and our conscious awareness of the measurement outcome**. But where exactly?

It may occur in various components such as the detector screen, in the circuitry, in the computer, or in our retina. However, it is important to recall that all of these components are composed of atoms and particles. The Von Neumann chain, which connects the detector to the mind, consists then of a series of quantum objects aligned one after the other. The lack of a distinct boundary between the quantum and classical realms is referred to as the Measurement problem.

Obtaining an answer to this question is quite a challenge, akin to searching for the Holy Grail. Von Neumann held the belief that wavefunction collapse occurs precisely when one becomes consciously aware of the experiment's outcome.

Another notable figure in early quantum theory who shared his views was Eugene Wigner, a fellow Hungarian-American who happened to be his study partner at Princeton.

In its essence, the Von Neumann-Wigner Interpretation proposes that consciousness has a crucial influence on the behavior and results of quantum systems. In contrast to other interpretations that consider quantum particles to be in a superposition of states until they are examined, this interpretation holds that the wave function collapses and a particular result is determined by the act of measurement by a conscious observer.

This perspective suggests that consciousness plays an active role in the quantum realm, rather than simply observing from the sidelines. It suggests that the way we perceive the physical world is a consequence of the interplay between consciousness and quantum systems. Essentially, our consciousness has the power to impact the behavior and characteristics of quantum particles.

In 1961, Wigner proposed a thought experiment to support the significance of consciousness. *The Wigner's friend experiment* (as it's called) goes like this:

Imagine if a friend of yours were to conduct the well-known double slit experiment and you knew very little about quantum physics. In this instance, your friend is aware that the experiment has been conducted and a single electron has successfully reached the detector. However, you are not aware of the outcome while (obviously) your friend is.

Here is an extra step in our Von Neumann chain: the information regarding this quantum experiment must first pass through your friend's conscious awareness before it reaches your own consciousness.

In this ambiguous moment, right after the electron lands on the screen and before your friend informs you of the outcome over the phone, your friend's mind is in a state of superposition, encompassing all potential results of the experiment. Naturally, from your point of view.

And once your friend reveals the outcome of the experiment, all the brain-wave functions in your consciousness converge to a single experimental result.

And if you were to inquire about the experience of being in a state of superposition to your friend, the response (with a puzzled face) might be along the lines of: "What are you talking about, have you

smoked something?" The wave function collapsed immediately after the physical experiment was finished!
However, since no information had reached you, there is no way that the collapse could have occurred from your perspective.

This evidence displays a fascinating conflict - various observers have differing opinions on when the wavefunction collapses. It appears that the collapse of the wavefunction is not an immediate event, but rather takes place gradually over a period of time. From this perspective, the collapse is seen as an ongoing and evolving process that varies for each individual who observes it.
And when multiple observers are present, this process tends to be extended. This interpretation presents a fresh perspective on the conventional idea of a distinct, clearly defined moment of collapse, and instead introduces the notion of a more intricate and fluid process.
Wigner believed that the experience of consciousness must have a role in both the generation and progression of the wavefunction collapse. However, there were other physicists who also raised doubts about the connection between the mind and the measurement problem.

While Bohr agreed with the Copenhagen interpretation, he was cautious about attributing any direct involvement of the conscious mind in this phenomenon. He staunchly defended himself against Einstein's accusation of introducing mysticism into physics.

However, Heisenberg had a more receptive attitude towards mystical interpretations and the potential impact of consciousness, particularly in the early stages. And even Schrödinger asserts that awareness is required for the meaning of physical reality in his 1958 lectures *Mind and Matter*.

With the renowned figures of quantum physics inclined to speak in "mystical" terms, it's no wonder that this concept has endured and gradually transformed into a more popular and spiritualized interpretation over time.

In the 1970s, books in the likes of "*The Tao of Physics*" and "*The Dancing Wu Li Masters*" made connections between Eastern mystical traditions and quantum physics, helping pioneer the genre. Although on the surface it may seem like a pleasant and harmless idea, the reality is that these works have unleashed an avalanche of consequences.

There has been a surge in self-help books and documentaries that make various claims, such as the ability to influence reality through acts of will, manipulate the outcome of events, or even shape the formation of snowflakes. Some even promise the potential to secure a promotion in the workplace! Or that the universe was created by our minds alone, and that there is no objective reality outside of our heads, and so on.

It appears that the only logical explanation for the uniformity of experiment findings amongst witnesses is that reality and experiment outcomes are independent of individual observers. Individual observers naturally have different backgrounds, stories, wishes, and minds. Sure, we could delightfully discuss the intriguing concept of a global consciousness influencing the collapse of a universal wave function. However, it's important to note that such discussions won't grant you any supernatural abilities to make quantum wishes come true.

Modern quantum theory has made significant progress since its foundation, although the Measurement problem still lacks a full consensus.

There are indeed precise explanations for why the wave function appears to collapse and conscious observation may have a role to play, but it's not quite what you'd expect.

Decoherence Instead of Consciousness?

The illusive line that separates the quantum and classical worlds is known as the Heisenberg cut but more and more physicists are of the opinion that consciousness and even measurements don't cause wavefunction collapse (at least not directly) - in fact probably there's not even a Heisenberg cut. It's possible that the collapse is just an illusion, and the various outcomes represented by the wavefunction could potentially continue indefinitely.

What?

The thing is: why do we lose sight of these multiple alternatives on our macroscopic size while witnessing them materialize on the quantum scale?

Some physicists believe that the solution can be found in quantum decoherence.

The alternative way to look at it is to imagine that the time-dependent wave function represents every conceivable history for the object. As a quantum system evolves, its "histories" split to encompass all conceivable futures permitted by physics. However, those divided histories have the ability to combine as well as to split.

Calculating the probability of a system transitioning from one state to another involves considering all potential paths that could result in that specific final state. However, for this to be effective, it is crucial that these alternative "histories", or branches of the wave function, stay coherent.

When decoherence takes place, the separate branches become permanently disconnected from each other, resulting in the loss of our ability to observe branches beyond our own. Let's get more into the meaning of coherence and decoherence.

In wave mechanics, coherence refers to the matching frequency and shape of a set of waves, as well as a constant phase difference between them.

In general wave mechanics, we say that a set of waves are coherent when they match in frequency, when the shape of the waves is the same, and when there's a constant phase difference between them (i.e. the peaks and troughs are either lined up exactly or have a constant offset). We notice this on the screen of the double-slit experiment - the zones where the particle/wave phases coincide we will see the bands of light appear on the screen.

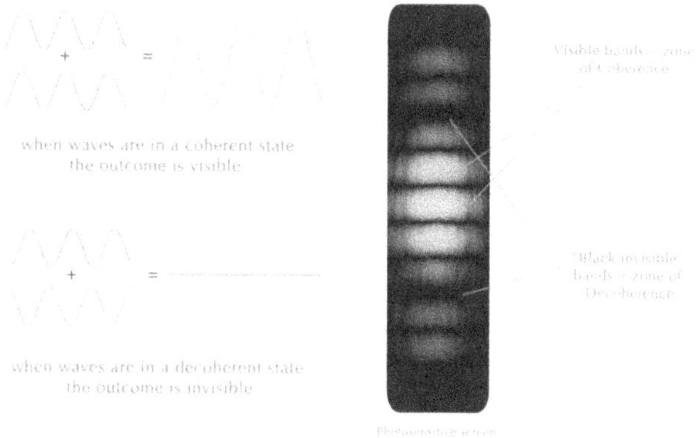

So, as observers, we are able to see only that part of the "history" where the phases coincide, instead, the other parts of the wave, or "potential histories" as we may call it, are not perceived by us.

To destroy coherence, all we have to do is screw up that phase relation.
As an example, suppose you were to fill one of the slits with a bunch of particles; this would disrupt the portion of the wave function that corresponds to a potential photon path. And thus, the

wavefunction that emerges from that particular slit acquires a random phase offset in comparison to the other slit.
Ultimately, what we would observe is a blurred image resulting from the convergence of multiple patterns, rather than the distinct alternating bands of light and dark that were visible in a state of coherence. It's important to note that the information about the relative phase was lost, resulting in the inability to observe the interference patterns.
We have become unable to differentiate the impact of various historical events. From our perspective, it seems that the wave function has lost its coherence and has undergone a process of decoherence.
And this is the reason why the interference pattern is destroyed when attempting to observe which slit the photon passes through.

Any measurement device must introduce some level of decoherence to the wave function before it reaches the screen. According to the decoherence hypothesis, it's not some magical effect whereby the wavefunction knows that it has been observed and so collapses. It's something else entirely.

Let's now take a moment to further enhance the level of intricacy. Let's zoom in on the exact moment when the photon reaches the screen. At that point, the photon stimulates the electrons in a pixel on the screen, causing an electrical signal to travel through wires to a computer and ultimately to our brain.

At some point in this process, the photon wave function might merge with the wavefunctions of the quantum particles in this chain. It is fascinating to consider the various potential paths that could unfold, where electrons are both excited and non-excited on the screen, and signals from these pixels travel to the observer's brain in superposition. But now the wave function is getting pretty messy and confusing.

There will be differences in the positions and energy of the electrons in the detector and the circuits. Phase differences are introduced between the various branches of the wave function as it becomes more intricate.

Ultimately, that expanding wavefunction includes the circuitry of the computer, and then the circuitry of our brain. There may still be multiple alternate histories propagating from the original double slit wavefunction, but by now each of those wavefunction branches corresponds to a specific configuration of matter and information in the computer and in our brain.

That specific wave configuration will lead to conscious awareness that aligns with a specific branch of the wave function, which corresponds to a singular location for the double-slit experiment. At this point as far as you're concerned, the wavefunction has collapsed and decoherence has taken place, according to your perspective.

However, it is possible that the initial double-slit wavefunction could continue to grow and become more intricate as it interacts with the wavefunction of the surrounding universe. That being said, we are unable to view its further branches at this time. Therefore, you shouldn't imagine yourself to be this all-seeing God-eye observer who can bring the wave function crashing down.

Rather you are embedded with the wavefunction and you see a slice of it, a slice corresponding to a single "history".

To conduct quantum experiments, it is crucial to isolate a specific portion of the global wave function and ensure its coherence. It is essential to possess knowledge about the relative phases across the relevant parts of the wavefunction. But for macroscopic scales, it's simply not feasible, it's inherently impossible.

Any contact with the external environment causes the phase information to leak into that environment.

And when I say environment, I'm referring to anything that isn't as meticulously controlled as your minuscule, isolated wave function fragment. And all of this, without delving into additional factors that can complicate experiments, such as entanglement and entropy.

It is also important to note that this paradigm for decoherence is not universally recognized, but it is gaining more and more backing. It is not widely accepted that decoherence provides a complete explanation for the measurement problem and wavefunction collapse.

However, it does have significance within the framework of the "*many words interpretation*" of quantum mechanics. Understanding why we lose the ability to perceive these alternative histories is crucial when considering their interaction with the macroscopic world.

CONCLUSION

Throughout this book, we have explored various concepts and theories that, certainly have sparked your curiosity. We have delved into the fascinating world of quantum physics and its intriguing nature.

Quantum physics, with its intricate principles and mind-bending theories, captivates our imagination and leaves us in awe about the mysteries it tries to unravel. The study of quantum mechanics has revealed the fundamental nature of the universe and challenges our conventional understanding of reality.

Its concepts, such as superposition, entanglement, and wave-particle duality, push the boundaries of our perception and push us to question the very fabric of existence.

As we conclude this journey, let us celebrate the captivating nature of quantum physics and its ability to ignite our snoopiness.

Navigating through these chapters we have taken a little but remarkable trek into the world of quantum physics, revealing the incredible and spectacular aspects that the subatomic world has to offer. The exploration of this field has probably not only expanded your understanding but has also challenged your preconceived notions of matter.

Furthermore, as we have seen, quantum physics goes beyond its theoretical wonders and finds practical applications in various fields. The potential for advancements in technology, such as quantum computing and quantum cryptography, is immense. These applications have the power to reshape our future and bring about significant changes in how we approach scientific and technological innovations.

Above all, this work serves as an ode to curiosity and scientific knowledge. It highlights the importance of nurturing our innate questioning, as it leads us to explore uncharted territories. The wonders of quantum physics remind us of the boundless beauty and complexity that exist in the unknown.

So, may the wonders of quantum physics continue to inspire and intrigue you, reminding you of the infinite possibilities that lie within our reach.

And whether you're a young student embarking on your scientific journey or someone seeking to reignite the passion for physics, embrace its wonders, keep the spirit of inquiry alive, and don't let your brain become food for zombies!

Oh, wait!

If you liked what you read and the way this material was conveyed, please consider leaving an honest review on the platform where you found it like Amazon, or referring this book to someone who might benefit from or enjoy its content, I'd really appreciate it.

Thank you!

And don't miss this one! I'm sure that you will be interested:

Whether this is your first time delving into it or you simply need a refresher course, <u>THIS IS THE MOST HIGHLY DIGESTIBLE FOOD FOR YOUR BRAIN AVAILABLE, right here!</u>

Are you looking for a way to learn the basic principles of physics in the easiest way possible? Many students will agree if you tell them that Physics is a challenging subject. It's true! From understanding Newton's Law to exploring how energy works, to how electric and magnetic fields are intertwined, Physics is a complicated subject. The problem with many books about Physics is that they are **often outdated and it makes learning the subject even more difficult.** Many resources that are used in the school as students' learning material are written long before the students were born.

The problem is not with **YOU** but with the resources you use to learn. **You need a book that is written on the same level of your understanding.**

Luckily for you, this ***Basic Physics for Beginners*** is the ultimate guidebook you'll ever need to learn the basics of Physics! This **comprehensive yet easy-to-follow book** on the principles of Physics will help you master the subject the easiest way! Here's what you'll get:

- **Comprehensive Topics on Physics:** From the history of the basic principles to the application of the rules and logic, this book covers everything you need to know!
- **Simplified Explanations on Basic Concepts:** With the fundamental principles explained straightforwardly, you'll no longer worry about your learning process!
- **Simplified Formulas:** No alien alphabets in this book, you are assured. The little Formulas in here are explained in an easy-peasy way! Don't panic. In short: highly digestible brain food!
- And so much **more!**

It's about time for you to **change your misconceptions about Physics.** With this guidebook, becoming a self-taught scientist is no longer a struggle! What are you waiting for? GET YOUR COPY NOW

Avoid the stargazing pitfalls and learn everything you need to know about astronomy without the complicated mumbo jumbo! - this guide makes astronomy so simple and entertaining that even a child will love it

Interested in the universe? Our solar system? How about stargazing and astronomy? If you're nodding your head, then ***Astronomy 101*** is the book for you! It can be really intimidating to try and learn about astronomy if you're a beginner. There's so much information out there, and it can be hard to know where to start. Not to mention, so many books out there go into heavy mathematics and physics equations that are just too complicated and boring. It's almost as if there aren't any books written for casual stargazers who just want to learn about pretty constellations.

But not anymore! This introductory guide will motivate and engage you in this fascinating subject. Written in a conversational style, this book contains definitions and explanations about the universe, our solar system, and recommended tools for observation - but always in a laid-back and easy-to-understand manner. We'll also show you how to stargaze on your own, without expensive equipment or classes. With our easy-to follow instructions, you'll be able to explore the night sky like a pro in no time!

www.ingramcontent.com/pod-product-compliance
Lightning Source LLC
LaVergne TN
LVHW022215200926
848481LV00006B/1983
* 9 7 9 8 8 9 8 4 3 3 5 4 3 *